HARD MAPLE, HARD WORK

by John Gagnon

Photography by Don Waatti

Drawings by Sandy Slater

A publication of
The Northern Michigan University Press
in cooperation with the
College of Arts and Sciences, 1996

© John G. Gagnon 1996

Northern Michigan University Press
1401 Presque Isle Avenue
Marquette, Michigan 49855

Library of Congress Cataloguing
Published 1996
Author : John G. Gagnon
ISBN 0-918616-17-4

To the people in this book—

For their encouragement and help with the manuscript I thank Timo Koskinen, Paul Lehmberg, Marilyn Monette, Don Kilpela, Bob Davis, Leonard Heldreth, Ron Johnson, Jim Carter, Beverly Matherne, Donna Hiltunen, Steve Hirst, Sue Ann Salo—and Mike Marsden, who got everyone blowing on the same coal.

TABLE OF CONTENTS

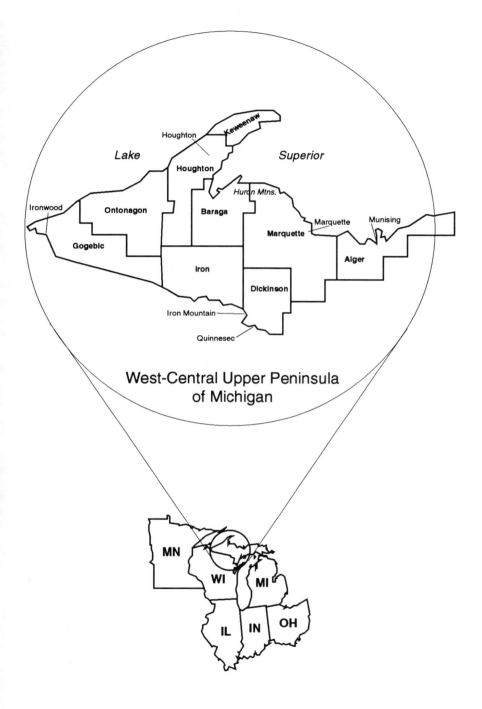

West-Central Upper Peninsula
of Michigan

LIST OF ILLUSTRATIONS

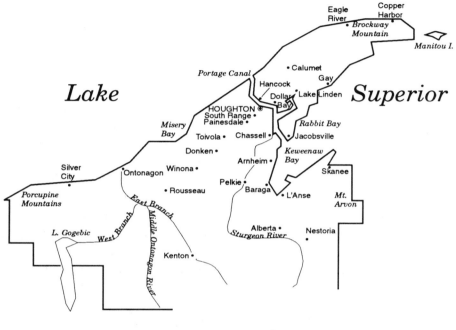

Upper Michigan's Copper Country

PROLOGUE

Along the south shore of Lake Superior, Michigan's Keweenaw Peninsula, like a giant breakwater, juts eighty miles northeast into the big lake. A rocky spine runs up the length of the peninsula; it holds a lode of copper, renowned worldwide, that sparked the first mining boom in the U. S. five years before the California Gold Rush. Those boisterous and adventurous days have gone the way of the pick axe and lunch pail, but the peninsula these days brags about another bounty: its superb hard maple. Some lumbermen and lumberjacks who've been around and know what they're talking about say northern Michigan hard maple is the best in the tree's North American range. Indeed, the Keweenaw lures lumber buyers from all over the world in the same way that certain French provinces entice wine connoisseurs.

I grew up in the Keweenaw in the 1940s and 1950s, but familiarity can breed oversight, so for many years I failed to see its natural beauty and wealth. My first impression amounted only to something I took for granted—lots of snow, which for months on end filled my boyhood days with yeoman's work. I sometimes had to shovel snowbanks that were fifteen feet long and taller than I was to make room in front of our house for holiday visitors to park their cars. During those years, I heard my first apocryphal story about Keweenaw winters, a tale in which a tourist stops at a gas station and asks the attendant, "What do people around here do in the summer?" The attendant replies, "Well, if it falls on a Sunday, we go on a picnic." Perhaps I was a dull-witted lad, but I believed that joke was a particularly clever illustration of a dubious distinction: namely, that the Keweenaw gets more snow than any other non-mountainous region of the U. S. The record is nearly four hundred inches. Snow and shoveling, then, were my first inkling that the Keweenaw might be a singular place.

That impression, however, faded during twenty years away from the area, an absence that exposed me to the natural beauty of other parts of the country. These different regions, as I saw them for the first time, all had a marked character—whether the stark, sculpted rim rock country of the Southwest with its bluffs and mesas, the stretching high plains of the West with countless hawks drifting and watching, or the monsoon-like downpours of the South with their verdant smells of humus.

But, upon my return home in 1980, the Keweenaw, seen for the second time, also began to appear exceptional. Here were alder-choked creeks and beaver dams with dark-backed, orange-bellied speckled trout, frost-encrusted trees haloed in the moonlight, resplendent fall forests, and Lake Superior, which one person has called "an unsalted sea." This renewed impression was underscored for me by an Ohio philosophy professor named Robert Long, whom I didn't know but had once heard speak. Long had lived for a time in the Keweenaw, and he described it as "land's end," an almost mystical meeting of "a hard land . . . and a cold and northern sea." When land and water so meet, Long said, "the land is more than the land, and the sea is more than the sea."

That image of the Keweenaw was brighter and more intriguing than the overriding one I had had of prolonged winters that tended toward dreary skies and dreary spirits. Those slowly passing dark months and moods are not easy to escape, for by continental U. S. standards, the peninsula is remote—"behind God's back," my old aunt says. In winter, by far the longest season of the year, travel out of the peninsula is often chancy, always lengthy. Copper Harbor, the northernmost town on the peninsula and in Michigan, is closer to Minneapolis, Milwaukee, and Chicago than to its own state capital of Lansing. Detroit is closer to Washington, DC, than to Copper Harbor. Three centuries ago, a Frenchman called the peninsula "the fag end of creation." Today a tee shirt reads, "Not the end of the world, but you can see it from here."

Maybe in summer you can. In winter, though, a low, gray sky shrouds the peninsula and bedevils the sun, which often is either shut out for long days or appears behind the clouds as a muted white glow that casts little brightness and no cheer.

In other respects, the peninsula is benign. Nothing in the forest endangers people but their own foolishness. Black bear encountered by surprise are always headed in the other direction, in a hurry, and no snakes or scorpions inflict unpleasantries. As my friend "Wimpy" Salmi says, "It's nothing like the Southwest where everything that crawls, it bites"—though in the summer a scourge of mosquitoes, black flies, no-see-ums, and deer flies vies with the winter for notoriety. Neither are there the tornadoes, hurricanes, floods, earthquakes, or mud slides that regularly plague life and limb in other parts of the country. "We get clobbered with snow," my father says, "but at least we know it's coming."

So it's a matter of shovel, shovel, shovel—often with a refinement of a shovel that is a local trademark: a three-sided, roughly three-foot-square aluminum scoop, with handles projecting diagonally backwards. Used to collect and move snow, the scoop works like a front-end loader that doesn't lift and that is powered by muscle instead of motor. This decidedly regional implement recalls another apocryphal story about the Keweenaw: natives threaten to leave the area with snow scoops atop their cars, head south, and settle in the first place where somebody asks about the scoop, "What's that?"—ignorance in this case amounting to snowless bliss.

The rigors of the Keweenaw's winters and the remoteness of the area mean it is sparsely populated. Although the peninsula has the same area as one Delaware and two Rhode

Islands, fewer than fifty thousand people live on this spur of land, which is comprised of Keweenaw County, most of Houghton County, and small parts of Ontonagon and Baraga counties. Keweenaw County, which forms the northeastern tip of the peninsula, is the least populated county east of the Mississippi River, and the Keweenaw County town of Eagle River, with a population of less than one hundred, is the smallest county seat in the U. S. Houghton and Hancock, with twelve thousand people between them, are the biggest towns on the peninsula. They sit across from each other on the steep banks of Portage Lake, which cuts a curly basin right across the midsection of the peninsula and connects to Superior at both ends. A dozen miles farther north sits the old mining center of Calumet, population nine hundred. At the height of the copper industry, Calumet Township had more people than all of Houghton County does now. Most of those original settlers were immigrants who brought their ethnic ways with them: the Finns their saunas, the English their pasties, and the French their *tuques*, which on the peninsula are called "chukes."

I once asked two people, a native and a passerby, to give me their first impressions of the Keweenaw. "A cemetery with lights," the passerby said. He meant the mining scars that blight the landscape: rusted shafthouses, dilapidated buildings, broken foundations, poor rock piles, slag piles, grey-black mine tailings called stamp sand that fill in lakes, and tall smokestacks that hover over ruins like crows over roadkill. The native had a completely opposite impression: "Blue, green, and white," he said, meaning water, forest, and snow, which the Keweenaw has in profusion.

The forest, which locals call the bush, covers nearly 90 percent of the peninsula. The dominant species of tree is hard maple, also known as sugar maple. Like its copper, the Keweenaw's hard maple is acclaimed throughout the world for its quality. Some lumbermen and foresters suggest that the exceptional maple results from Lake Superior, which either embraces the peninsula with mild temperatures or clutches the peninsula with heavy snows. Whatever the reason, Robert Long's "land's end" mystique apparently has worked its magic.

Some time after I moved back to the Keweenaw, an old lumberjack showed me a small bow saw that he had cut firewood with for eighteen years. "If that saw could talk," he said, "she'd have something to say." So, too, I imagined, with Keweenaw hard maple, and I searched for its story in this land of the north. I found people who fashion a tree, and a tree that fashions a people.

Fred Aho

THE OLD DAYS

Like an eagle looking for an updraft, Fred Aho chased trees all his life and soared to prominence among lumbermen. Before he was old enough to shave, Fred skidded logs out of the bush with horses. He ended his logging days sixty-six years later. In those years between ages fourteen and eighty, Fred became known in America and Europe as a specialist in *figured* timber—trees with unusual and handsome patterns in the grain, like bird's-eye maple, a rare and hard-to-spot wood, the best of which is found almost exclusively in northern Michigan.

The first time we meet, Fred tells me that someone once asked him: "How can you tell bird's-eye maple just by looking?"

"Just by looking," Fred answered.

Fred lives in Hancock, midway up the Keweenaw Peninsula. Hancock is the only town in the U. S. with street signs in both English and Finnish. On the cold, early-winter day that I first drive to Fred's home, tongues of wind-blown snow lick the road, and clouds of snow, like wraiths, swirl off of the tops of three-foot-high banks. Fred's house is welcome, warm refuge from the storm.

Right off, Fred shows me around. His place is richly paneled with hardwoods that are finished with clear sealer except for dark stain on the black walnut in the living room. This mistake was made many years before by the carpenter who installed it, and to this day Fred is irked. Staining walnut apparently is like painting a flower. "May as well have put up tarpaper," Fred says so disgustedly that I cringe for the carpenter who goofed up the job, for if Fred is still vexed, he must have been as angry as the devil when it first happened.

1

Fred is Finnish. His parents were among the many nineteenth century Finnish immigrants who settled in northern Michigan and brought with them a reputation for untiring work habits, a spunky way of making do, and a language barrier more severe than other immigrants faced—a situation suggested even today by such family names as Honkavaara, Kujansuu, Tuoriniemi, and Waliuaara.

Fred took the best and the worst that his heritage gave him to make his way in the world. He is small and lean, about five-foot-eight, with thick white hair, a florid complexion, and heavy-lidded eyes that sometimes just squint, especially when he laughs. He gestures a lot, draws pictures, and makes notes to illustrate what he's saying. He is excitable. "So-so-so," he stammers in his hurry to get his story out. He is friendly, but one night when I visit and turn my tape recorder on, he barks, "What are all these questions about?" He points to my tape recorder. "Just what are you after with that thing?" It takes me a half hour to calm him down; the whole time I fear that he will banish me from his home, as he would that carpenter of long ago. But Fred displays that old man's paranoia only once and from then on is gracious and cooperative—and modest about a reputation that stretched all the way from northern Michigan to London, England.

Fred was born in 1903. When he was young, his parents, nomad-like, moved with the timber: wherever it was being logged, they lived. When Fred was fourteen, the family lived forty miles southwest of Houghton in Rousseau, which is more of a location than a settlement. It was the fifth place that Fred moved to as a child. At the time, his father ran a small lumber camp of about twenty men, and even then young Fred contributed to the operation. He did his father's bookkeeping, his letters, and his payroll, and he skidded logs out of the bush—with a single horse because he was too small to handle the heavier gear for a team.

Early in the winter of his fifteenth year, Fred toiled at his father's camp with only summer clothes to wear. He got cold and disgusted and decided to quit woods work. He put the horse in the barn, had enough of a sense of responsibility to feed it, and then walked home fourteen miles. He told his mother he would go back to school the next day. Around midnight he heard his mother and father quarreling. That fight shaped the rest of his life. "Mother wanted me to go to school," he remembers. "Father said, 'I don't know why he has to go to school. He does just as good a job as a man, and I have to pay a man thirty-five dollars a month and board, and I don't have to pay him anything.'" His father won the argument; Fred, better clothed but somewhat chastened, returned to the bush.

"So-so-so, I'll show you a picture of my first winter in the camp," he says, as he slides off the kitchen stool and shuffles off with short steps to the bedroom to get a picture. In it are two boys in a man's world, Fred and his younger brother standing with fifteen men in a snow bank in front of the bunkhouse. "I'm highly educated," Fred says. "I got one day in the eighth grade."

More than seventy years later, dressed in heavy wool pants, suspenders, and a plaid wool shirt buttoned tight to the neck, Fred looks as though he can still feel the cold from that first winter in the bush. His demeanor is just as warm as his clothes—and as excitable as March weather. He is no longer agile but still burns with energy despite his years. When I visit him one evening, he has spent three days shoveling snow off the roof of his summer home.

Fred's father's one concession that first winter in camp was that Fred didn't have to work outdoors in the snow. Instead, Fred became the cook's helper. In subsequent years, though, he worked outside and, while yet young, he learned much about what powered most woods work: horses. Fred's father, besides logging, also traded in horses, but he had absolutely no horse sense. "He would hit a horse in the head with a cant hook, anything," Fred says. "I wouldn't let him drive my horses to water." By contrast, Fred became a good horseman. He attributes that to his mother's kind ways. "She used to say, 'Don't hit your horses like pa does.'" As Fred grew, so did his affection for the animals. "I love horses like I do a person," he says. "If you try to get more out of 'em than what's in 'em, then it's your fault." He says one of his great delights in life was a saddle horse he once had that "rode like a rocking chair."

Fred used horses so much that he came to have an uncanny way with them. He remembers a time when he visited a sawmill in southern Wisconsin. The sawmill owner's wife raised horses. Fred had trained mostly workhorses and was delighted to see thoroughbreds. He fussed over a white Arabian with black spots. He picked up the horse's feet and tapped on the hooves. He always did that when shoeing so that he could get the horses used to the hammering. Then he pulled the Arabian's ears and mane, rubbed his nose, "talked sweet" to it, and gave it a sugar lump. "I always carried sugar lumps," he says. When Fred started to walk away, the Arabian grabbed the back of his shirt with its teeth and followed Fred to the gate. Fred walked it back to the stall, talked to it, and fussed over it some more. Three times he walked to the gate; three times the horse followed him. Finally, Fred walked backwards and the horse stayed put.

"What'd you do to that horse?" the owner asked Fred.

"My father used to be in the horse business and I used to train 'em," Fred tried to explain.

"I been around horses all my life but I never saw anybody, a stranger, go to that horse and he wouldn't want you to go," the owner said.

He offered Fred a job and said, "You could ask whatever you want and my wife would pay you."

"No, I'm not that kind," Fred answered. "I been in the timber woods all my life."

"Who taught you to train horses?" the man asked.

4

"The horses," Fred said.

Being trained by what he bossed was a lifelong pattern for him, like a mentor being taught by a student. And, just as the horses taught him how to train horses, lumberjacks taught Fred how to boss lumberjacks. "The lumberjacks been good to me," he says. "They helped me in every way." He had a knack for getting along with the woodsmen. In the early days, he says, typically sawyers were Finns, teamsters were French, railroad men were Swedes (It was said of them: "Give a Swede a can of chew and a bottle of whiskey, and he'll build a railroad to hell and back—or at least halfway"). The Poles, Fred says, were "this and that and every other thing."

At an early age, Fred showed common sense about working with all these people; he knew enough not to act their superior even though he was the boss's son. "Most of the fellas in those days, they couldn't speak to these workers in a nice way," Fred says, "where, with me, I felt as though I was one of 'em. I had a feeling among these fellas. It's God's blessing. I had no problem with these different nationalities. I'd say, 'Why are you doing this?' and 'Why are you doing that?' and they'd explain it to me. I kinda learned to be a woods foreman before I was twenty-one years old."

Fred's father's lumber camp didn't log in the summer when the sap runs in the trees, for if the trees were cut then and the logs left piled, the sap stained the wood dark. "The color wasn't true" is Fred's way of putting it. The maple was being used mostly for furniture, and the buyers didn't want it stained. Only big operations like Ford Motor Co., which had dry kilns to pull the sap out of the wood, cut all summer. Small operations like Fred's dad's cut only in the fall and winter. The time to start felling was when the leaves started falling—the sure sign, Fred says, that the sap has stopped running.

All the while he worked for his father, then, Fred spent summers on a road gang, which proved invaluable to his

5

future as a lumberman. Cars were becoming common in the Upper Peninsula, and crews were widening roads to accommodate them. Fred's summer job was to hold the surveyor's stick for the civil engineer behind the surveyor's instrument. One day Fred, always yearning to learn, said: "Say, will you teach me how to run that machine? I'm pretty good in figures." The surveyor agreed to teach Fred. The trick was to establish a benchmark and determine where and how much to cut, and where and how much to fill in the roadbed. "That was nothing for me," Fred says, for despite not having finished grade school, he took to numbers like he did to horses, and he could figure in his head faster than most people could figure on paper. "It's God's blessing," he says.

When Fred was in his mid-twenties, he and his father, who had a violent temper, got into a "chewing match." Fred quit and went to see Pat Gallagher, the superintendent, or what Fred calls "the walking boss," of Von Platten-Fox Co., which ran a big logging operation in the western Upper Peninsula. Gallagher hired Fred as a saw boss, but after about a month the bookkeeper quit, and Fred wanted that job. Gallagher gave it to Fred, who was content for about eighteen months.

Then Gallagher had to build a railroad in the bush to haul logs to the company's mill in Iron Mountain, which is far south of the Keweenaw, near the Wisconsin border. Fred, ever bold and hungry for opportunity, asked Gallagher to let him oversee the building of the railroad. His experience on the road crew and his ability in arithmetic were enough to sway Gallagher, so Fred again got another job that he coveted. Once completed, Fred's railroad extended more than twenty miles and had four locomotives that served Von Platten-Fox's four camps—each of which bunked, fed, and worked two hundred men cutting many sections of timber and loading ten railroad cars a day. "That's pret'near a half a million feet of logs a day," Fred says proudly. "And I built the railroads—surveyed 'em and built 'em." An eighth-grade

6

dropout managing the tough calculations and planning gives him a pride that seems not boastful, but earned.

Fred stayed with Von Platten-Fox until the late 1930s. Then a man from France came to the area to buy bird's-eye maple. Fred had sent maple burls to France because their grain is so crazy and complicated that they don't split and are ideal for pipe bowls, but the interest in bird's-eye was new. Bird's-eye grows throughout the hard maple in the Upper Peninsula. On the stump, the figure has no obvious tell-tale feature, at least to the inexpert estimation. Sometimes the eyes are evident in the bark, but usually not. Fred says he can spot it by what he calls a "corset"—a slight tapering, and then flaring back to normal, low on the trunk. Fred says the Upper Peninsula has the finest bird's-eye he's ever seen—the finest maple in general—especially in the early days when loggers were making the "mother cut" on the virgin hardwood.

The first time he tells me about the Frenchman looking for bird's-eye, Fred gets up from his kitchen stool and beckons me to follow. He leads the way to the bedroom hallway. There, outside a room paneled with different hardwoods native to the Upper Peninsula, is bird's-eye maple used for door trim, doorstop, and other molding. Putting bird's-eye to such use employs a patrician wood for a plebeian purpose, like using pearls for worry beads. But that doesn't cloud Fred's appreciative eye, and he lingers over one eight-inch-wide board that has bird's-eye almost as thick as a good case of freckles. His fingers caress the wood softly. "There isn't maple any place to compare to this maple," he says.

Fred attributes the tree's superior quality to heavy snows. Parts of the Keweenaw get what is called "lake effect" snow. Dry, arctic air from Canada takes on Lake Superior's warmer moisture, hits the forested land of the Keweenaw that slopes up, cools as it rises, then drops the moisture as snow. Even though foresters say hard maple thrives in the cold and prefers a short growing season, Fred believes these snows act as a ground insulator, and the tree roots don't get so cold that the tree goes fully dormant during the winter. "As soon as the snow is gone, you see buds on these trees here," Fred says—before they appear farther south.

The Frenchman looking for bird's-eye spotted Fred's ability to identify the figure at a glance, gave Fred two hundred dollars, and told him to be in Montreal on a certain date. Fred went to see his boss, Gallagher, who was in his sixties, and told him that he admired him but didn't want to end up like him: working Monday through Saturday in the camps and seeing his family only on Sunday. Fred says he told Gallagher, "I want to get around, to see electric light once in a while." Gallagher understood, gave Fred his blessing, and Fred set off for Montreal. The journey was the start of a pursuit that would last the rest of his working days, for from that date forward Fred became a figured timber buyer.

From Montreal, he took the Canadian Pacific northwest to the end of the railway at Mont Laurier, Quebec. He inquired there about where to find another figured timber that the Frenchman wanted—curly birch. Further north, he was told. Fred hired a horse and cutter and traveled another twenty-five miles to Lac Saint-Paul, a small settlement near timberline. There, big but short, stood curly birch. Fred couldn't spot it like bird's-eye. "It looks like anything at first," he says. "I had to take the bark off." He soon learned to tell the tree by the bark, which he says looks a little bit stretched and has a slight glaze to it.

"They have a beautiful curl," he says of the birch. "They have a longer curl than on the maple here—like a marcel wave in a woman's hair."

"What's a marcel wave?" I ask.

"You ain't been around women much," Fred says.

Then he gets off his stool, shuffles purposefully into the living room, and points to some door trim made from curly maple, which he says is similar to curly birch. It has quarter-inch-wide, three-inch-long, wavy, blond bands that run at right angles to the grain. They look like the haphazard, sinuous forms of trout schooled up on a spawning bed.

The locals in Canada were happy to find somebody to buy the stuff. Fred traveled by horse and sleigh to small logging operations and sawmills and bought all the curly birch he could find. He stayed two winters, buying only logs that had been cut when they were frozen because, as with maple, the sap discolored the birch in the summer. "I gave 'em good money," Fred says. "I gave 'em more for logs than they could get for lumber. I got along with those Frenchmen pretty well. I was like a little god with those guys, you know, when I give 'em a fifty-dollar bill or a hundred-dollar bill. I paid everything cash."

If he was a well-heeled man in spartan country, he also was a man with a religion but without a church—a Finnish-Lutheran in French-Catholic country—and Fred started going to Mass with the French loggers every Sunday. His adaptability was by then as characteristic as his curiosity. Fred also got by in the language. He could read and write in English and Finnish, and he taught himself French by writing the French phonetically in Finnish because Finnish has no silent letters. "I learned it quick," he says.

After his sojourn in search of curly birch, Fred's reputation was international, and, a few years after he returned to the Upper Peninsula, he received a phone call from England. People there wanted quilted maple. Could he find some? Fred had never seen quilted maple, but he knew there wasn't any in the Upper Peninsula or he'd know about it. Quilted maple is different from curly maple: the grain runs in all directions, looking like patches on a quilt. Fred says he thought about where to look for it and told himself: "There's a lot of maple in the foothills of the Appalachian Mountains, and that's where there's coal. Here, there's maple and copper; there, there's maple and coal. I wonder, I wonder . . ."

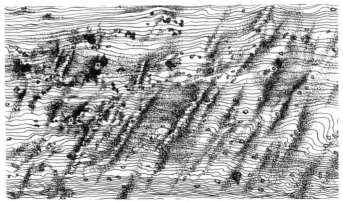

bird's-eye maple

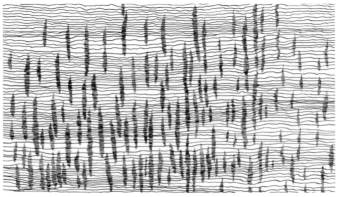

curly maple

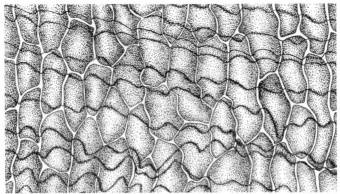

quilted maple

10

He went to Pennsylvania and found it there. He tries to explain his hunch, but it's like trying to describe a song to a deaf person; he can't do it and says only, "It's God's blessing. It's something told me where to go. That's all I can say. I don't say I know. I took chances. But it comes to me natural. I was among trees all my life. That's why I always had a job, because I knew my timber."

When he found quilted maple, once again he had to take the bark off to know if he had it. He soon learned how to spot it because the trunk had a little bulge to it, like an oaken barrel. He axed a sample slab off a tree, sent it to England, and shortly received a telegram telling him to buy as much as he could get. Fred supervised the logging. He was real fussy: no cant hooks on the logs, no chains wrapped around them, just tongs in the ends. He loaded up several train cars and sent them to the East Coast. From there, the logs were shipped to England and put to a regal use appropriate for an ornate wood: they became some of the woodwork on the H. M. S. Queen Mary. "They say it's just beautiful," Fred says.

He walks a bit shakily to the living room, tells me to follow, opens a drawer in an end table, rummages around, and gives me a piece of quilted maple about as thick as a balsa wing on a model airplane and about as big as a postcard. It used to be three feet long, he says, but he kept whittling off pieces to give away samples. The piece he gives me is his last. The raised, patchwork grain makes the wood look wrinkled, like the skin of a prune or raisin. It's a striking wood, but I feel as though I'm taking Fred's last dime— worse yet, his last evidence of a cherished memory. But Fred insists, and I take the wood.

After the quilted maple venture, Fred headed back to the Upper Peninsula. He built many logging railroads and was a woods foreman across northern Michigan and northern Wisconsin. He came to know those forests as well as his own name, and he also bought timber in Montana, Iowa, Illinois, Missouri, Wisconsin, and Canada.

Fred worked until he was eighty and for the last forty years also bought veneer for plywood manufacturers. For veneer, he looked for logs with no defects, no knots, no blemishes. It's tricky stuff to spot, he says, but he used his tried-and-true strategy of relying on the experts to tell him how to please experts, in this case the factory men to tell him how to please factory men. "They would tell me and show me what's good, what isn't good, and why," he says. He emphasizes again and again that he was just one of the boys, not a know-it-all. "I didn't learn all this by myself," he says. "Somebody's got to show you." But one thing was certain, Fred says: "They knew I was listening."

Fred talks enthusiastically about his working days, but he is obviously self-conscious. He reminds me of a little girl I once saw do magic tricks for a group of adults. When she stood on the carpet in the entryway to the living room, her father said, "Come closer." "I can't," she said, "I'll fall off the stage." Similarly, Fred is mindful of being the center of attention. It makes him uneasy, and he says, "I been reminiscing about stuff that was past tense twenty-five years ago. I just gave it to you as I grew up with it. I didn't know anything about it when I started, and I don't know a helluva lot now. If my head was as smart as my feet, I'd be a smart man. It's a very picturesque life, but it never come to me that I should expound on it."

I ask Fred whether he was happy at his work. It is a stupid question: only somebody who grew up after World War II and who had the luxury of a choice of careers would ask a question like that of an old, old person who was pulled out of school after one day in the eighth grade.

Happy? "I didn't know any better, put it that way," Fred says. "I never gave it a thought."

I ask him if he ever resented being forced out of school and into the woods.

"I didn't have time," he says.

I ask him if he misses the old days.

12

He answers without nostalgia: "Now there's no more camps. It was quite an ordeal then. Now everybody got an automobile, and you can drive an automobile right to the tree almost. But I tell you, you know, it's like everything else, like everything else. The old Model T days—you wouldn't drive a Model T anymore. So that's about the way I'd explain it. The Model T days are over."

We return to the hallway by the bedroom with the eight-inch beauty of a bird's-eye. Fred runs a hand over it, gently, as though he were touching glass or a pretty girl's face. "I got some beautiful bird's-eye here," he says, meaning a twenty-mile swath of central Houghton County between the two towns of Painesdale and Winona. "I cruised that timber and I marked the trees, but I never cut any bird's-eye in the summer. You see, that sap in there would turn black and it wouldn't be right. But when the leaves dropped, boy, I tell you, I got some beautiful bird's-eye between Winona and Painesdale. Oh, geez, it was beautiful."

We linger in the hallway. Time and familiarity have not diminished his love of good wood. He rubs the bird's-eye some more, talks of his searches for it, and his voice is as warm as a summer pond.

Swede Intermill

MORE OLD DAYS

W. W. "Swede" Intermill says that saunas made Finns like Fred Aho the cleanest lumberjacks in the normally lousy logging camps that dotted the forests of the U. P. in the first half of this century. The communal baths were routine for the men—and death for lice, for the Finns used to "cook" their lousy clothes on the bath stoves, and oldtimers say that outside, in the change room, they could hear the lice popping like miniature fireworks. Swede, who is English and Swiss—he got his nickname when he was a towhead kid—once took a sauna with some Finn woods workers and says it took a week to cool down and another week for the red flush to fade from his face.

Saunas and warmth are on Swede's mind one chilly Fourth of July, when a steady drizzle drops from a dismal sky. The weather forecast assured Keweenaw residents of sun for the holiday.

"It's tough to predict the weather being next to Lake Superior," Swede says. "She makes the weather no matter what the weatherman says."

"We need the rain for the bananas," somebody else says.

Swede's bearing is the opposite of the weather. He has a warm smile and guileless blue eyes. He and I are to ride in three Keweenaw parades this day, and together we amble from a parking area to our float. Tall and thin, he walks with a slight stoop. I slow my pace to match his, for he is old and his step is hesitant, almost fragile, as though a rush of wind could stop him cold or knock him down. His tentative walk belies a strong, sure sense of matters, especially of northern Michigan hard maple, and I have joined him to learn what he knows about the tree.

15

The float we are on commemorates the Keweenaw's industries, and Swede is dressed as a lumberjack: leather-laced boots, pants rolled up to just below the knee, plaid shirt, suspenders, and a hard hat that he takes off now and then to reveal thick, white, wavy hair. I ask Swede why his pants are rolled up so high. Years ago, he says, lumberjacks cut the legs off below the knee so their pants wouldn't drag in the snow and get wet and heavy. "They used to say you could find out how deep the snow was by finding out how high the pants were cut off," Swede says.

He was born in "aught-nine" in Fort Dodge, Iowa, ninety miles north of Des Moines. Early in life he had a link to the northwoods, for when he was young he saw smoke from big forest fires in Michigan, Wisconsin, and Minnesota drift all the way down to Iowa and "shut out the sun." Swede's grandfather often took him into the Iowa woods, which the youngster enjoyed so much that he decided at the age of twelve to be a forester. The first in his extended family to attend college, Swede studied at Iowa State University, graduated with nine other foresters, and then worked for the U. S. Forest Service at the Laona Ranger District of the Nicolet National Forest in Wisconsin. Shortly after World War II, Swede hired on with Consolidated Paper Company of Wisconsin Rapids, Wisconsin. He supervised the company's Upper Peninsula operations, which were headquartered at the small settlement of Donken in north-central Houghton County. When Swede retired from that job, he and a partner teamed up and did woods work on their own until 1984. Swede's career in the woods industry lasted more than fifty years, so his presence lends authenticity to our Fourth of July float.

Our parades are ten to twenty-five miles northeast of Donken—in the Keweenaw towns of Lake Linden, Dollar Bay, and South Range—but Swede is close to his former workplace in spirit because he carries a broadaxe. A broadaxe, with a blade about fourteen inches long, has one side

16

curved and one side flat. Lumberjacks used a broadaxe to hew timbers. The axman would strike the log parallel to its length. As the axe penetrated the wood, the flat side sliced cleanly into the meat of the log, and the curved side, which faced the perimeter of the log, wedged a slab outward and split it off. In this way, a round log could be hewed almost square.

Swede recalls seeing a team of Canadian axmen hew pine logs for the military during World War II. The Navy needed big timbers to plug up holes in the hulls of damaged ships so they could limp into port. The Canadian axmen put on a show with broadaxes that still fascinates Swede today. They laid a chalk line on the logs, and, swinging the axe, grunted when they hit the log "like these tennis players on TV when they hit the ball." The axmen made that chalk line disappear as though they wielded erasers, not axes. Swede says the men were artists, and their pride was evident; the leader of the team worked alongside the rest of the crew but stood out because he wore a derby and white silk gloves.

Chain saws have largely replaced axes, and I tell Swede that a friend of mine uses two of them to notch logs for a cabin. Even with power saws, I remark, the work is dreadfully slow and painstaking. "It's a lot of work to do a good job," Swede reaffirms, "but you really got something when you're through with it. It's as tough a job as a woman putting clothes together. Patience—that's the art of it."

Our first parade is in Lake Linden, one of three old milltowns on Torch Lake. A mile away is my hometown of Hubbell, which has so many mining ruins that some young boys I know call it Rubble. As we move slowly down Lake Linden's main street, Swede periodically takes a swipe at my bald head with the broadaxe and tells bystanders he shaved my head with it—a joke that due to his kindly bearing doesn't get tedious. He calls me "boy" a lot. His tone is not demeaning, but comradely. Besides, I am half his age and a young buck in his old eyes.

Our second parade is in Dollar Bay, a small town on a cove off of Portage Lake. It has only a few hundred residents, but it boasts two sawmills and a maple flooring company. The latter industry prompts Swede to talk about hard maple's hardness. Some people, he says, believe oak is harder than maple. But, Swede says, rub the edge of a half-dollar real hard across a piece of oak, and it'll leave an indent—not so on hard maple.

For our third and last parade in South Range, eight miles southwest of Houghton, we gather at a small woodlot on the edge of town, where big pines stand high above hard maple saplings. The mix of trees triggers a lesson by Swede on sylvan matters. One way to classify trees is by whether they like sun or shade, he explains. Maples are a tolerant species, which means they tolerate shade. Pines are an intolerant species, which means they don't tolerate shade but need lots of sunlight to grow. A maple can drop thousands of seed pods in a sunny, warm spot, but none will open up and take root. A pine can drop thousands of cones in the shade, and none will release seeds. "Put a pine cone in an oven and you'll see how much it likes warmth," Swede says. "It'll open up like a flower." Pine needs an open, sunny, warm spot for a start, and maple needs a canopy of trees to provide shade for a start.

What tree comes up under another, or what tree fills in an open area, is called tree succession, Swede says. In the northern Upper Peninsula, pine and poplar follow fire and a windfall, what Swede calls a "wind throw." Beneath cover, maple grows and eventually succeeds the trees that give it shade. Hard maple, then, is a climax forest because it provides the shade for its own kind to succeed it. Once hard maple gets a toehold in a forest, like a dominant gene, it begets only more of its own.

Swede then talks about people—he calls them "bughouse environmentalists"—who don't want timber logged. "The same guy that kicks about cutting timber when it's mature, he will thin his radishes so he'll get good radishes, but you're not supposed to cut a tree," he complains. Surveys today show as much timber in the U. S. as there was in 1900, he says. Still, he adds, trying to convince some people about the benefits of logging is like "trying to train a coyote to be a Labrador." Those people don't understand that the forest always recovers from human intrusion, he says. "Old Mother Nature always got that ace in the hole. She's a tough old girl." Swede says talking about cutting timber is like talking about religion. "People get fanatic. It's sure funny." He shakes his head in bewilderment, but his chagrin stops far short of rancor.

And the weather this day stops short of severe. It could be worse, for I—and others I know—remember a Fourth years ago when it snowed lightly. This day, though, is just cold, and the nip in the air doesn't deter Swede from his gregarious ways. He mills about before and after each parade and chats with people. He seems to know everybody and has a cheery word for them all. After the South Range parade, he compliments an old friend about not having any wrinkles.

"They're all inside," she laughs.

"That's a fooler, then," says Swede.

Another woman asks him how he is.

"Kicking up a little dust, no gravel," Swede answers.

I don't see Swede for quite some time after that, but I hear about him twice. One acquaintance recounts Swede's saying, "I'm the kind of guy still amazed that you can flip a switch here and a light goes on over there." Another acquaintance who meets him says that, despite his age, Swede has a handshake as hard as an ironwood tree. The two reports make me determined to talk to Swede some more, but it is the following spring before we get together for several interviews.

19

Our first is at Swede's home on Portage Lake, just outside Chassell, which is eight miles south of Houghton. I arrive upset at being upbraided by a colleague for a mistake at work, and I tell Swede about it. "Don't worry," he reassures me. "Some people like to pick the fly specks out of the pepper." Right away I like him more.

Swede's house, like Fred Aho's, was built by somebody who loves wood. One room is paneled in ash, another in elm, the kitchen in yellow birch, and the bedrooms in bird's-eye maple. Throughout the house are several bowls carved out of hard maple burls. When Swede finishes showing me the place, we go to the living room where big bay windows look out upon Swede's backyard, a stretch of Half Moon Beach, and Chassell Bay, which is dimpled by a light wind. Swede's backyard is notable for many trees and many, many bird feeders. Never in a hurry, Swede pauses to watch dozens of birds flit from feeder to feeder and also eat seeds that have dropped on lingering patches of snow.

"Isn't that a gang of 'em!" Swede says.

There are goldfinches, chickadees, and purple finches.

"I don't know why they call them purple finches," Swede says. "They're more raspberry."

We watch them for long minutes, then abandon the birds to their paradise and retreat to the couch and nostalgia. Swede is dressed in wool pants, a green plaid shirt, and green suspenders with the words "Logging Congress" running down them in yellow. He never tires of my questions, and he talks about the old days with understated eagerness. When he talks, he rarely gestures. He just sits, barely stirring, like an old languid river. His deliberate, soft voice sometimes trails off to almost a whisper. Time passes agreeably in his company.

For two hours on our first visit, we look at pictures of Swede's World War II days. In college, he joined the ROTC program—a big inducement was the extra set of clothes that the military gave him. When World War II started, he was assigned to a forestry outfit, and spent thirty-two months on Guadalcanal in the southwest Pacific fighting thorny "wait-a-minute" vines to log mahogany. His work mates were "young and full of fire," he says, and it was good duty because "you put your rank in your pocket." Hearing him talk about that infamous island as the site of a routine, agreeable logging operation during wartime strikes me as ironic—like enjoying seeing all the relatives at a funeral.

After the war, Swede returned to the Forest Service in Laona for a short while, then took Consolidated Paper Company's job in the Upper Peninsula. His memories of those old logging days paint a vivid cameo of camp life: of "the jacks" sleeping in two-man bunks with a "snortin' pole"—about three or four inches in diameter—between them so one man's thrashing didn't disturb his bunk mate; of the only bug dope that woodsmen knew—balsam pitch; of oldtimers getting dibs on the cooler-sleeping bottom bunks; of bum cooks who got drunk on lemon extract. Good cooks were hard to find, and Swede hated the job of looking for one.

Swede especially wants to talk about the old sawyers and their two-man saws, which were powered by muscle and sweat and worked to a rhythm as regular as the beat of a two-step. He can't find the words to express his admiration of those sawyers. "Geez, they could saw," he says, "Oh, man." Once he met a sawyer who used to rig up an inner tube to a stake, tie one end of the saw to the inner tube, and when he'd pull, the stretched rubber would pull the saw back. "Best partner I ever had," the fellow told Swede. "He never complains."

The lumberjacks called the two-man saws "come-to-me, go-from-you's," and a sawyer, Swede says, was as good as his sawdust. Actually, sawdust is a misnomer; if it was dust, the sawyer wasn't good or the saw wasn't sharp. A good sawyer with a sharp saw turned out long strings of shavings called "worms." Other marks of the trade: a poor sawyer would misnotch a tree, causing it to split as it fell and leaving a "barber chair," a big splinter standing upright on one side of the stump; a poor sawyer would drop a tree up against a standing tree; a poor sawyer would leave "widow-makers," branches from felled trees that broke off and were left hanging in standing trees, ready to fall and hurt or kill someone; a poor sawyer felled trees in a tangle and made it hard on the teamsters skidding the logs out of the bush.

One tangle Swede remembers was in the winter of the late 1940s when he had to supply his company's mills with more pulpwood. To get the job done, he hired eighty miners who were on strike. The move was a mismatch, like hiring a preacher for a sideshow. Everybody in the bush got their dander up over it. "Of course, a lumberjack would never admit a miner could get the job right," Swede says, "and the teamsters cried and complained about the logs laying crossways and not being cut in an orderly fashion. And the miners said, 'Boy, you go into a mine, and it's either a hot mine or a cold mine, a dry mine or a wet mine, but you know how to dress every day because you have the same conditions all year. Out here, one day it's so hot you sweat, and the next day it's so cold you can't live with it.' So they weren't very contented either. But they were desperate. They'd been off work for a couple of years. They were bitching, but they weren't bitching at you because they were glad to get work. But they'd never done this, and they hated it. We just had a awful time."

The experience reinforced Swede's admiration for the expert sawyers, and again and again he struggles to express his feelings for them. His imagination can't find the right words; he may as well try to catch a moonbeam. "Geez," he says, "some of those fellas, I tell you, they could just really saw."

Swede's crews cut part of a big stretch of forest that runs for nearly fifty miles from Houghton west to Ontonagon, and from Lake Superior inland ten to fifteen miles—an expanse of land drained by eight rivers running north into

Lake Superior: the Ontonagon on the west, the Salmon Trout on the east; between them, the Graveraet, the Elm, the Misery, the Sleeping, the Firesteel, and the Flint. The terrain is lumpy, with hill upon hill folding into each other like a stack of potatoes. Swede's men cut hardwood—mostly maple, birch, and basswood—and some softwood—mostly hemlock—in this part of north-central and northwestern Houghton County. The hemlock grew in the drainages and the hardwood on the ridges. Swede supervised three crews logging more than twenty sections, a total of almost fourteen thousand acres, from after World War II until 1973. For fifteen years during the same period, he also supervised the cutting of poplar on the Baraga Plains, about fifteen to twenty miles southeast of his Donken headquarters. And he supervised some work much farther southeast in the Upper Peninsula. He didn't like logging outside of Houghton County, though. He explains that the phrase "Snow-on-the-road" meant a lumberjack had grey hair, and what put snow on his road was logging south of the Keweenaw Peninsula where he encountered "cold weather, bum timber compared to over here, steep hills, rocks, and bad swamps."

At the height of the Misery Bay operation, Swede supervised one hundred men in the fall and winter and two hundred men in the spring and early summer. The extra crew in spring and summer was needed to peel the bark off of hemlock logs. The hemlock bark was sold to tanneries all over the U. S., the hemlock logs were sent to Wisconsin Rapids to Consolidated's paper mill, and the hardwood was sold to another logging company that had bought the hardwood timber rights in the Misery Bay area. That company in turn sold all the hemlock and other softwoods it logged to Swede's operation. Swede admires the men who peeled the bark off of hemlock nearly as much as the old sawyers. He says the best peeler he ever knew was a fella called the Million Dollar Kid, so named because he sewed his money into patches on his pants.

Swede oversaw one of the last logging camps in the Upper Peninsula. Camp 27, located west of Donken, ran until the early 1960s. In the 1940s, the camp housed about forty or fifty of his big crew, but over the next twenty years the camp gradually got smaller due to the mechanization of the industry. First, chain saws thinned the ranks of sawyers needed. Then, big tractors and "big and awkward" skidders that Swede says looked like bulldozers without blades dragged logs out of the bush and eliminated the need for teamsters and horses. Then, trucks replaced even more of the teamsters and horses who worked the loading jammers. A jammer was a big A-frame. A cable was run from a team of horses, up over an A-frame that was high above a truck or a railroad car, and then down to the logs. When two men, called hookers, attached one end of the cable to the log, the horses on the other end pulled and raised the log up onto the truck or flat car. Both that horsepower and manpower were replaced by trucks with winches. Then, self-unloading trucks, with swiveling booms that lifted and loaded the logs, came along and put the hookers out of work. Finally, loggers drove in automobiles to work each day, so suddenly the camp became as obsolete as the two-man crosscut saw.

"There was just a terrific change," Swede says. Consolidated kept Camp 27 as long as it did partly out of a soft spot for some of the older lumberjacks. By the 1960s, Swede says, the camp was only a reminder of the days "when everything was done the tough way." Swede welcomed the changes and the progress, much of which came in the 1940s. When self-loading trucks came along during World War II, Swede says, "We thought that was just about the biggest advancement possible to make."

Although he likes to talk about the old days, Swede doesn't romanticize them the way bigger-than-life lore enshrines the white pine days of the late 1800s in Michigan and Wisconsin. Take, for instance, the old saying in the Upper Peninsula and northern Wisconsin that supposedly measured how much vice—in the form of whiskey, barroom brawls, and scarlet ladies—the miners and loggers indulged in when they visited two northern Wisconsin towns, Hayward and Hurley. It was said that a person could go from "Hayward to Hurley to Hell"—I'm not certain whether Hell was the town in southern Michigan, or that other place. Whichever, Swede has misgivings about that slogan and the reputation it suggests, at least for his day. He says that every year while the camp was open he drove lumberjacks to Hurley in the spring and fall when the woods was too wet to work in.

Listen:

"About the first of May, why, I'd make a trip over to Hurley and see a couple of madames over there that we knew and that we'd get our good men from. And with all that they say about Hurley, a lot of that was bunko, you know—excitement. These fellas, they were all single guys who'd go over to Hurley to stay during the spring breakup. A madame that knew what she was doing, well, the guys would come in, give her their pay, and she'd take out her book—they called that the book—and write down what it was. And he spent off of that. And if he treated somebody to a drink, which wasn't near as often as he liked to be treated, she would take it off of that, and she'd dock him for board and room. And so I'd tell her, 'I'll be over on the Thursday or Friday after the first Monday in June to pick up a crew.' I wanted them for hemlock bark peeling on the Donken landing. And she said she'd have them all set. We'd come over on that day, they'd be all sober, they would have clean clothes on, they would have a pack sack of clean clothes, and everybody would bid her goodbye. And we'd have two or three cars of men, seven or eight to a car. And the only thing we had to do, when we closed camp in the fall, we'd have to take them back to that same place."

Swede adds that the reputation of Hurley "was for the big city dudes who came up there and didn't know how to handle themselves. The lumberjacks and miners, they weren't so crazy. Maybe in the years long before that, when the only outlet there was was to go in there and try to fight, but I didn't see that. There was no trouble."

Swede loves hardwood logging, if it's done properly. One telltale feature of logging, he says, is what's left behind. Clearcutting leaves behind virtually nothing. What Swede calls a commercial clearcut, or even-age management, leaves behind just small trees. What Swede calls selective logging, or uneven-age management, leaves behind several generations of young trees to replace old, mature trees that are harvested. Swede recalls with fondness the year 1937 when he was with the U. S. Forest Service in Laona, Wisconsin. He and two other foresters planned and supervised the thinning of a hardwood stand in the district. A few years ago the three of them revisited the stand. It had been cut three times since that first thinning; each time 25–30 percent of the stand was harvested. Today it's ready for a fourth cut. "You can see many stems, and they're all different ages," Swede says. "That's what good hardwood forestry is all about. Seeing that hardwood stand—that gave us about as big a thrill as anything."

But his efforts weren't always that rewarding. Like the misdeeds of a wayward son worrying his mother, Swede's Misery Bay operation was an unremitting heartache for him. He says Consolidated bought the timber rights from Calumet & Hecla Mining Co. "C&H," as the copper mining firm was known, wanted all the hardwood cut down to twelve inches and all the softwood down to ten inches—a typical commercial clearcut that harvests almost everything of any size and value. Swede wanted to take out only the largest trees and leave the rest for future loggers. "We met with them, we argued with them, we talked to them," he says. "We showed them figures. All they said was, 'All you have to do is cut and get the hell out of there.' We just begged to selectively cut. But it was their land. I don't know how many times we went up there and talked to them and talked to them and talked to them. I tell you, it would get to me."

Swede is not a profane man. Usually his exclamations are "Holy man" or "Holy mackinaw." But the bad memory of that heavy cutting so vexes him twenty-five years later that he is roused to curse. "Boy, when you cut it like that, god-damn, you could shoot a cannon across it and never touch anything," he says. One old saying for such heedless logging is: "We cut it so heavy that the woodpeckers had to carry lunch boxes." But Swede is disinclined to couch an unfortunate situation in humor.

"Did it all make you sick?" I ask.

"Oh, did it ever," he says, raising his hand in a gesture that is emphatic for its infrequence.

Swede says his operation left some trees behind because Misery River country is being logged again more than two decades later. "So you can see," he says, "we didn't scrape it too tight. We didn't take every damn thing that was there, but we took way too much. If we could have just been able to manage it—just take the big stuff out. All they said was, 'Don't bother us. Just cut it and turn it back to us.'"

Although he did as he was told and the memory is like a big knot in his throat, the experience didn't diminish his love of working in the bush. "Oh, I just loved it," he says of his labor. "I sure did. I really liked it. It was a lot of get-up-and-go, but if you had good timber, it was really fun to see those old trucks coming outta there. Boy, they used to come out of Misery Bay, I tell you, one after another." He pauses and continues with obvious sadness. "But we sure failed. We failed in not being able to convince them to cut for uneven age."

Swede says the Misery Bay country is known for its bird's-eye, and I ask him how good Keweenaw hard maple in general is. He says that a big lumberman from Wisconsin always told him that the best hard maple in the Midwest comes from Houghton County. Whereas Fred Aho links the area's good maple to snow cover, Swede believes the weather on the Keweenaw Peninsula makes the maple so good. Houghton County doesn't get the really cold temperatures that areas farther south get because Lake Superior, which seldom freezes over entirely, has a warming effect on the peninsula in the winter, and a cooling effect on the peninsula in the summer. Lake Superior modifies the temperature "something terrific," Swede says, and the temperatures, pretty mild for north country, make for good maple.

I ask Swede about the saying that "hard weather makes good timber."

"I don't think that's quite the truth," he says, and points out that hemlock gets what he calls "shaky." The term describes how the tree, in severe weather, splits apart along the growth rings. Swede says that condition is common in hemlock in the Misery Bay area, where "there's real hard weather, cold and wind."

"What about maple?" I ask.

"It'll make frost cracks down the maple tree," he says. "It'll be a defect. It goes clear through to the heart, and it'll be stained and dark."

In 1972, Consolidated finished logging in Houghton County, and the company wanted Swede to transfer to either Wisconsin or Minnesota. But Swede and his wife had found a home, so Swede chose to retire rather than move. His leisure status was short-lived because his wife tired of his hanging around the house. "I married you for better or for worse, but not for lunch, so get going," she told him just two weeks later. Swede decided to fulfill a lifelong dream. "I kind of had it in the back of my noggin that I would like to have a sawmill before I closed it up," he says. He and his partner bought a hardwood mill in Kenton, in southern Houghton County. Swede was the field man, out in the bush buying logs from loggers, but he used to hang around the mill just to see nice logs turned into beautiful lumber, and to hear the high-pitched whine of the saw. To him it was a song.

But a discordant note sounded a year and a half later: lightning struck the mill and it burned down. "That was about the saddest thing that ever happened to me," Swede says. He and his partner decided to rebuild. "When that old girl got built and it ran, I think that was about as big a thrill as I've got out of anything."

The pleasant memory of that time in his life endures as strongly as the climax maple forest. Indeed, the forest came to mean peace of mind to Swede during one of the frustrations of this period in his life: bidding on timber sales conducted by the U. S. Forest Service. The federal agency's foresters estimated how much timber a stand would yield, then invited loggers to bid on the rights to cut it. "You can't go through and count all the trees," Swede says. "You have to just take samples. You had the acreage, so you'd decide if the stand was going to underrun the estimate or overrun it. Then you made your bid. You didn't get paid according to what you cut. You got paid by what they said was there and what you bid. That was tough on everybody, and sometimes up to eighteen or twenty guys were bidding against you. If the timber is real good, you want it so bad, but you can't get wild and bid high and lose your shirt, you know, so those timber sales were a tough job. You gotta talk to yourself." Every time he bid on one, Swede would go for a walk alone among the trees and do his figuring. In this way, the bush became more than trees, logs, and lumber; it became a place for sorting through confusion, a place for good, long thinking, a place, says Swede, "where I used to be sure my mind was clear."

Swede retired for the second time in 1984 at age seventy-five. Logging northern Michigan's forests made for a good life, he says. "I never wanted to do anything else." Trees have taught him the patience that exudes from his quiet, almost imperturbable bearing. "There's no use getting excited," he says, "because our crop isn't every year, it's every sixty years."

He has seen what such patience can do. When he lived in Laona before the war, he had men from the Civilian Conservation Corps plant a two-foot-tall white spruce in the yard of the ranger's headquarters. In 1979, forty-four years later, the spruce was cut down for the national capitol's Christmas tree. It had grown more than a foot a year in height and stood fifty-nine feet tall and had a branch spread on the bottom of thirty-three feet. It was perfectly symmetrical because it had no competition for sunlight.

That tree stood alone. Those in the bush occasionally stir Swede to be philosophical. He sees parallels, for instance, between forest trees and young people, and he says, "I tell the kids, you know, you grow up just like a young tree. Your neighbor trees decide whether that tree will lean this way or that way, or if it's going to grow straight. The more shade they throw on that tree, it leans a little this way and grows crooked. And with good neighbors and friends, a kid will grow straight."

Yes, Swede says, with both the earnestness of youth and the conviction of old age, "I like trees. And there's nothing prettier than a nice maple tree."

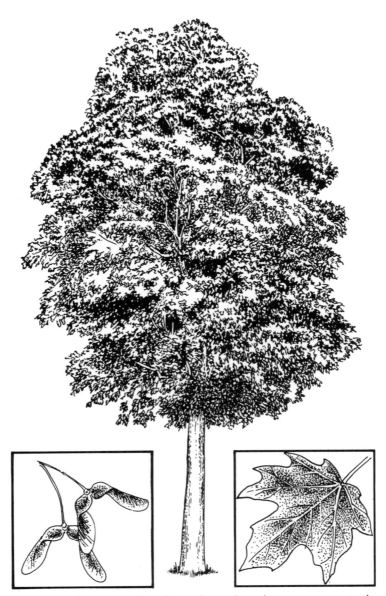

acer saccharum, the American hard or sugar maple

THE TREE

Old woodsmen say that Guy Nordine, who sold Swede In-termill his sawmill, once had a dog that bit only Republicans. Nordine says the story "is a joke in reverse," because those oldtimers definitely know he's a Republican and actually would like to have a dog that bit only Democrats, if there were such an animal. Guy himself is good-natured about this teasing, and he says of all old woods workers, regardless of their politics: "Now we log in our sleep and dreams." But if tuckered-out old men rest their muscles except in reverie, steeled young men flex theirs except in wind and rain and mud, and they continue the tradition of hardwood logging in Michigan's expansive northwoods.

Those forests sit atop Michigan like a crown: a soft russet in the budding spring, green and lush in the summer, garish in the frosty fall, and grey and somber in the winter. An aes-thetic wonder and an economic boon, they annually attract sightseers, tourists, hunters, fishermen, bikers, snowmobil-ers, campers, and cross-country skiers, not to mention log-gers. Michigan has more than seventeen million acres of for-est, the most in the Midwest, and the biggest stands are in the western Upper Peninsula—an area that stretches roughly 130 miles from Marquette west to Ironwood, and 150 miles from Keweenaw Point south to the Wisconsin border. The western Upper Peninsula has nearly five million acres of forest—88 percent of the land in the region. These forests are mostly northern hardwoods: red oak, ash, aspen, basswood, paper or white birch, yellow birch, soft maple, and hard maple. The dominant tree is hard maple. Accord-ing to the official proceedings of a forestry conference at Michigan Technological University in Houghton, hard maple is "perhaps the single most successful species" of tree

33

in Michigan's north woods. With its frequent neighbor, yellow birch, hard maple comprises 54 percent of the trees in the western Upper Peninsula. That level of stocking means that the western Upper Peninsula has the "largest remaining commercial volumes" of hard maple in the state, according to the state's Department of Natural Resources.

Hard maple, or sugar maple, is one of about twelve hundred kinds of forest trees in the U. S. It is a hardwood, which is what foresters call a broadleaf tree, as opposed to a softwood, which generally bears needles. Maple, a deciduous tree, sheds its leaves each fall, as opposed to an evergreen, which generally keeps its needles all year. (In these classifications, potential confusion abounds. All hardwoods are not hard; some softwoods are harder than hardwoods. Soft maple, so called because it is softer than hard maple, is a hardwood like hard maple.)

Hard maple, an adaptable tree, can thrive where temperature extremes reach as high as one hundred degrees and as low as minus-forty degrees Fahrenheit. Like a splayed hand, the leaf of a hard maple usually has five dominant points, but can have more. Its seeds, called *samaras*, are winged. In its range, hard maple tends to dominate hardwood forests. The reasons, like the points of its leaf, number five. It is fecund, hardy, opportunistic, tough, and tolerant—traits that enable it to grow widely, albeit slowly.

Its fecundity is easily measured. Its plentiful, far-roving seed can be carried by the wind up to a thousand feet. A hard maple forest in northern Michigan was calculated in one study to release more than eight million seeds per acre. Another study in northern Michigan showed that a hard maple forest sent seventy thousand seeds per acre into a surrounding ten-acre tract. Hard maple seeds sprout up to a million seedlings per acre. These seedlings constitute survival of the species through sheer volume; for, according to Dr. Glenn Mroz, a Michigan Tech forester, much "carnage" occurs in the forest to deplete the stock: only one hundred of those one million seedlings will become mature trees.

The attrition would be greater except that hard maple stands up well in the harsh forest environment. The tree is as hardy as it is hard. The seedlings bounce back well, for instance, from damage inflicted by browsing deer. "No other northern tree species can withstand such heavy browsing," one researcher notes.

Hard maple is both an early bird and a tough old bird. The seedlings begin growth in the spring before other species do, thereby getting a jump on the competition, and the seedlings are resistant to natural mortality and catastrophe, such as fire. One forester describes hard maple as "the asbestos forest" because of its tolerance to fire, but Mroz suspects that the moist maple forest generally precludes fires. Another forester told me that in the early part of this century, fire fighters would drive forest fires into the hardwood stands, where they would die out.

The main reason that hard maple tends to dominate other hardwoods in its range, though, hinges on patterns of sunlight and shade, two criteria that foresters use to classify maples and other trees. Swede Intermill talked about tolerant and intolerant species, noting that hard maple is a tolerant species because it survives in the shade. Research has shown that seedlings readily establish themselves as an understory beneath a canopy of trees casting shadow on them. And the shadows can be deep, for hard maple seedlings need only 2 percent of full sunlight. They start in the shade and grow as little as a few millimeters a year, sometimes not at all. In the latter cases, Mroz says, seedlings "probably put on some kind of annual ring, but it's not big enough to measure if it's only one set of cells. It can grow real slow."

Foresters, then, commonly find hard maple seedlings that are eighteen inches high and thirty or more years old. Seedlings in such a state are suppressed. At the point when an overhead tree is removed, the seedlings get more sunlight, are released, and begin to grow at a normal rate, which is still slow—perhaps an inch in girth every five to ten years.

In one of several interviews with Mroz, I ask him whether more sunlight might cut short the suppressed stage and allow hard maple seedlings to grow more quickly sooner. Apparently, he says, hard maple needs shade for a while early on in its life. He says that one experiment in the Keweenaw is instructive: maple trees exposed to full sunlight are thirty years old and only three feet tall; by comparison, aspen, a northern hardwood with a sunny disposition, will grow to thirty feet in half that time.

Although hard maple is widespread compared to many other species, it otherwise sets no records. Hard maple can live three hundred to five hundred years, but doesn't have the antiquity of a sequoia, which can live ten times longer. Hard maple can reach a circumference of fifteen feet, but doesn't have the girth of, again, a sequoia, which is the biggest of trees and can have a circumference of one hundred feet. Hard maple grows about one hundred feet high at maturity but isn't as lofty as a redwood, which can grow four times higher. Hard maple is neither the lightest of woods (the corkbark tree of the Southwest is), nor the heaviest (south Florida's black ironwood, nearly twice as dense as hard maple, has that distinction). Hard maple usually has a pear-shaped canopy, not the sweeping, majestic crown of an oak. Hard maple needs clean air to survive and therefore isn't as common an ornamental tree in urban areas as an elm, which can survive in smog and soot.

In fact, hard maple is known more by its uses than by its appearance or name. Thus, as the eastern red cedar cleaves easily and makes good pencils, hard maple grows strong, withstands shock, and makes good bowling pins. As eastern white pine grows tall and straight and in sailing days made good ship masts, hard maple in a vertical position doesn't compress and makes good pillars. As black walnut is shock-resistant, takes a nice finish, and makes nice gun stocks, hard maple is stiff and strong, and makes good ladder rungs.

Although such uses, like its appearance, are strictly run-of-the-mill, hard maple has spawned an entire industry. Trees have done that throughout history. Hemlock bark, which provides tannin to toughen leather, attracted shoemakers to Pennsylvania in the early days of this country. Similarly, in Vermont, hard maple sustains the maple sugar and syrup industry. In the Keweenaw, in the rest of the Upper Peninsula, and in northern Wisconsin, hard maple gave rise to the nation's flooring industry. For maple is "strong, tough, hard, and heavy," one wood merchant notes, and its premier use is in floors. Maple is good for bowling pins, bobbins, croquet balls and mallets, billiard cues, furniture, hockey sticks, veneer, railroad ties, paper, shoe trees, pianos, guitar necks; all these uses aside, though, hard maple's properties make it ideal for floors. It doesn't splinter, it doesn't abrade, and "it wears forever," says John Hamar, former president of Horner Flooring Co., which is located in Dollar Bay and is the northernmost hardwood flooring company in the country.

Mere chance didn't dictate the location of Horner Flooring Co. The forests of the Keweenaw, of the western Upper Peninsula, and of northern Wisconsin have attracted every maple flooring company in the country—partly because of the area's abundance of hard maple and partly because of its location in the central U. S. Hamar says that no flooring companies are located in the East or New England because the freight costs they'd have to pay to market to the rest of the country would be prohibitive.

Under Hamar's leadership, Horner developed portable basketball floors that grace sports facilities from North America to Asia to Europe. In building this reputation, Hamar has bought hard maple from throughout the Great Lakes states; he says the best maple Horner gets is from northern Wisconsin and northern Michigan, and the peerless maple is from Houghton County.

Maple is the second wide-ranging species of tree in Michigan history to earn a notable reputation. In the second half of the nineteenth century, the state's white pine stands were coveted. At the time, lumbermen said the Michigan, Wisconsin, and Minnesota white pine forests would last until the end of the world. In fact, they were virtually logged out in a few decades, and all that remains of them are the Paul Bunyan legends they birthed in their demise. After the white pine was exhausted, Michigan became one of the leading producers of hard maple in the United States, a position she still maintains, and lumbermen the world over cast an especially envious eye at Houghton County's hard maple—a wood of storied dimensions.

Take the testimony of Michael Lorence, market development specialist of hard maple products for Mead Corporation's Northern Hardwoods Division near South Range. Lorence has harvested trees, graded and bought logs, and graded and marketed lumber for everything from furniture to pallets. He says that Europeans and Asians come to the Keweenaw "looking for the best of the best hard maple." He adds, "We've got it."

The Keweenaw grows such superior hard maple that he recently encountered a situation that many salesmen only dream of: a virtual blank check for an order. A European buyer told Lorence that he would buy as much hard maple product as Northern Hardwoods could produce. "This gentleman is the top-of-the-line floor manufacturer in Europe," Lorence says. "He's big. He's world-renowned for excellence. So it's a perfect match because this area is world-renowned for its hard maple."

Lorence's job enables him to see hard maple from all over its range, which stretches from Tennessee to southeastern Canada, the mid-Atlantic states to Iowa, Maine to Minnesota. The Keweenaw, he attests, grows the best hard maple. "I'm not exaggerating. I can back that up," he affirms. "These Europeans can buy hard maple in Canada and on the East Coast, which is much closer to Europe and less expensive to ship out of. Yet they come two thousand miles farther inland to buy. They aren't paying more money just because they want to keep another source at hand. They're buying our wood because it's the best they can buy."

Forester Keith Brey, who has worked with trees for nearly twenty years, oversees nearly a quarter-million acres of hard-wood forest in the Keweenaw for Lake Superior Land Company, a subsidiary, headquartered in Calumet, of Champion International Corporation. He tells me, "The Japanese are very high on our maple. They come looking for it." He adds, "I've never come across maple as good as you find in Houghton County." He particularly likes the maple near Misery Bay in the northwest part of the county, where Swede Intermill used to log.

All these men agree that the best hard maple is located specifically in Houghton County, not elsewhere in the Keweenaw Peninsula. To the north, in Keweenaw County, the soil is rockier and oak does better. In fact, Keweenaw oak is as prized as Keweenaw hard maple, they say, but it's a minor species.

Why all the accolades for Houghton County maple? They arise from several factors. For one, Keweenaw maple is prized for its whiteness. "It's beautiful, it's white, it's gorgeous," Lorence says, adding, "I've been told in many countries in Europe that we have the whitest hard maple in the world."

Another factor that foresters cite about Houghton County maple is its small hearts. Lorence calls them "pencil hearts" because they're often virtually that small. Brey calls them "veneer buyer's hearts" because often a tag one inch square, used for singling out defect-free veneer logs, will cover the heart. Technically, says Mroz, the heartwood is bigger, but only a small portion of it is reddish-brown in color. Technicalities aside, Brey says that he's seen bigger maple in Wisconsin than in northern Michigan, but the dark hearts constitute half the diameter of the tree, and the sapwood isn't as white. Brey says that Europeans, as well as Koreans and Japanese, "are all looking for white wood—the whiter the better," and many come to the Keweenaw to get it.

The third and last factor that foresters agree on is that Keweenaw maple has little mineral stain, which causes elongated oval streaks in the wood that are darker than the sapwood and as telltale as a birthmark. Mineral stain is common to the west of Houghton County, foresters say.

If foresters agree on the quality of Keweenaw maple, they disagree on the reasons for it. The tree hides its secrets like a matron her age. Whereas Fred believes the key is snow cover, and Swede credits mild temperatures, Hamar tends to favor superior soils. Northern Houghton County has weather and snowfall similar to northern Alger County more than a hundred miles southeast, he says, but the maple from that area doesn't compare with the Keweenaw's.

Brey and Lorence think the reason is a combination of climatic factors: lake-effect moisture, mild temperatures, and good soils. "Our winters aren't typical to anywhere else in North America," Lorence says. Adds Brey: "It's much milder here than points south. The lake really tempers the extremes. You don't get that real bitter cold, and you don't get that extreme hot and dry weather. The trees here don't look anywhere near as stressed as they do farther south." Brey offers two examples of stresses: frost cracks in the trunk from severe winter cold and bud mortality in the upper branches because of late spring frosts.

Lorence points out that the hard maple in the Keweenaw has little competition. "Everywhere you look, there's hard maple," he says. "It comprises sixty-five percent of the forest that Mead owns." That figure jibes with federal and state

40

estimates of the extent of the maple-birch stands in the western Upper Peninsula.

These stands grow not only plain hard maple used for flooring, but also bird's-eye, that same beautiful configuration in hard maple that Fred Aho used to pursue. On the stump, the figure in bird's-eye consists of little bumps on the tree, beneath the bark. Swede calls them "miniature burls." On a board, these bumps are sanded smooth and become little swirls or circles that suggest eyes.

The eyes can be as tiny as the point of a pin or as large as the head of a spike. They can be as thick as the dots on a fawn or as scattered as feathers in the wind. They can occur all over a tree or on one side. They are found on two-inch-high saplings and on full-grown trees. The bird's-eye figure is a "distinctly North American" product, one expert says. Although the figure is found in other species of trees and in other parts of hard maple's range, bird's-eye predominates in hard maple in the Upper Peninsula, which has been called "bird's-eye maple heaven." Brey sees a lot of it, not all of good quality. "The farther north you go in Michigan, the more you find," he says. "It gets to the point where, as you get to Copper Harbor, every fifth tree has some kind of bird's-eye, so you're always looking for it."

No one knows how bird's-eye is formed, although several theories have been advanced: rotten heartwood, unhealthy or deformed trees, virus. Even though hard maple is a tolerant species for a while, one theory holds that bird's-eye occurs where a dense canopy of trees provides too much shade and results in slow or stunted growth, particularly during a tree's younger years. A recent study by Michigan Tech, Mroz says, supports that theory and suggests that bird's-eye seems to grow "slower than normal." Besides that lone clue, all that foresters know for certain is that planting seeds from bird's-eye doesn't produce bird's-eye. The solution is more complex and probably will involve a combination of genetics and environment, Mroz says. The Michigan Tech study also confirmed Fred Aho's knack for spotting it. Bird's-eye, Mroz says, has a barely detectable "Coke bottle" appearance, a slight bulging and then tapering of the trunk.

Besides mystery, secrecy shrouds bird's-eye. One buyer refuses to be interviewed and says that he and his workmates have a policy not to reveal anything that they know about bird's-eye or its whereabouts. The apparent reason is the price of the wood. Foresters say that a superior bird's-eye log routinely will sell for thirty, forty and fifty times more than plain hard maple. Mroz knows of three logs that went for six thousand dollars. Brey knows of single logs selling for three thousand dollars. With that kind of money at stake, bird's-eye buyers are "pretty tight-lipped," Mroz says, adding, "If people knew how much money bird's-eye is worth, they'd start rustling logs."

Lorence says bird's-eye constitutes only 1 percent of Northern Hardwoods' product, which totals about twenty million board feet a year. (A board foot is one inch thick, twelve inches wide, and twelve inches long.) One in a thousand bird's-eye logs is really good, Lorence says, and the prices are misleading. "Anybody can find one or two good logs and get top dollar," he explains, "but you still have to market the lower grades, too—the pallet and box material." Bird's-eye is rare, he says, in great demand, and, in comparison to the entire forest, it is only "the cherry on top of the sundae."

. . . Or the jewel in the forested crown of Michigan, for maple forests are a vital part of Michigan's landscape, recreation, and economy. Indeed, throughout history, trees have proven invaluable to all of humankind. A folk saying perhaps sums up that usefulness; trees, it says, "provide the wood of your cradle, and the shell of your coffin." Hard maple is no exception to that literal and symbolic worth; in Michigan, especially in the Keweenaw, the tree is the epitome of it.

Jim Johnson

THE MARKER

Swede Intermill believes that children and trees grow up with the same influences: good neighbors make good fiber—both moral and wood. By that standard, Jim Johnson, a forester and timber marker, is the steward of the forest; for in his job he culls the bad, nurtures the good, and harvests the big and bountiful. Down come the trees screwed into a twist. Down come the hollow, the rotten, and the crooked. Down come the trees in their prime. On the stump stay the young and promising, and also the healthy and middle-aged—they live for the axe of another day, another decade.

In his work, Jim Johnson routinely takes aim with his paint gun, pumps the trigger, and shoots an orange ribbon of paint through the air to lay a bright, ragged circle around a tree. Jim is marking timber—squirting the paint around each tree that should be cut down by loggers. In effect, he blazes what will be the sawdust trail that loggers will some day leave in the bush. And he is doing something that he has wanted to do ever since he was a boy: work in the woods. "I just loved the outdoors and I thought I'd get an outdoor job," he says of his decision to be a forester.

Jim was born in 1928. When he was young, he had three uncles who had farms in the Keweenaw, and he worked for them every summer. He took to the labor like a black bear to a blueberry patch, and by the time he was fourteen, he had decided to be a forester. When he graduated from high school, he even skipped taking the summer off and began his studies right away. "I was anxious," he says. "I wanted to get going. I didn't want to waste any time." He brings to his job today that same purposeful attitude.

Jim is marking a forty-acre, privately owned woodlot across Chassell Bay from Swede's home, and about a mile or so inland. The woodlot is fairly flat, and when the going is easy, as on this tract, Jim can mark about ten acres a day. He is on his second foray into this stand of hardwoods. His dog, Skila, a rare, wavy-coated black retriever, accompanies him. I am along to watch. It is after sunup. Thunder from the tail end of a night rain grumbles far away. The rain has made the woods wet and the ground quiet to walk on. Cobwebs on the undergrowth glisten with raindrops that, like bits of crystal, catch the morning sun filtering through the forest canopy.

As he works, Jim gives me a running commentary on what he's doing and why. He looks at a forked tree and sprays it. "No use growing a half-log tree," he says. He comes to an area where he marks all the trees; each has a defect of some sort. "No use letting junk grow," he says. He sprays a tree with a diseased spot where the tree is rotting. Around the rot is a callous or ridge, called a canker; the tree, like the human body, is trying to heal itself. "It looks just like a cobra's head," Jim points out. He sprays another tree, which has a horseshoe-shaped growth called a conk, another symptom of disease.

Jim is of medium height and build. Today he wears green work pants and shirt, a camouflage cap, boots, and a carpenter's apron that is besmirched with dirt and orange paint. In his pockets and vest he carries a can of paint with a spray attachment that can squirt the paint twenty feet, pliers, bug dope, a Boy Scout hatchet, and an extra can of paint. He also carries a Biltmore stick, a device that measures the diameter of a tree, and in his vest pocket are a tablet and pencil to tally the trees to be cut. He marks inferior trees of any size for pulp sticks to make paper and superior trees for saw logs to make lumber or veneer.

Before retiring in the late 1980s, Jim worked for thirty years at Michigan Tech's Ford Forestry Center in Alberta, about twelve miles south of L'Anse and Baraga, twin towns that flank the foot of Keweenaw Bay. While there, he supervised the management of a four-thousand-acre hardwood research forest, parcels of which he logged four times. Jim let three favorite hard maples in that forest grow to thirty-two inches in diameter, bragging size for northern Michigan maple. When Jim's management project was about to be terminated due to budget cuts, one of the last things he did before he retired was fell those big trees.

"Did it bother you to cut them?" I ask, wondering whether he might have been sentimental about them.

"It never bothered me," he says, thinking dispassionately. "I cut so many trees. You've got to cut trees to grow trees."

That viewpoint singles him out as a forest manager who often is caught midway between environmentalists, who generally want to preserve timber, and loggers, who, he says, tend to overcut timber. The middle road that Jim takes between these two extremes is the practice of selective logging that Swede yearned to do. Since his retirement, Jim spends many days marking timber on a contractual basis for private landowners. The woodlot that he is marking on the day I join him is a stand of red maple, red oak, basswood, and hard maple. Jim says that hard maple likes a well-drained, rich, moist soil, and grows well on higher land, especially north-facing slopes, which get less sun and therefore aren't as dry as south-facing slopes. Hard maple generally grows slowly but is adapted well to the climate of northern Michigan, which Jim calls "sixty-day corn country."

Jim explains that a hardwood forest should have four stories: big, mature trees that yield logs, pole-size growth beneath the big stuff, saplings beneath the poles, and seedlings from three inches to three feet tall beneath the saplings—all in all, three stories living under one leafy roof. That ideal distribution, which Jim calls "structure," ensures that "when you take a tree out, you've got something coming in its place." The woodlot that Jim is marking is stocked with mature trees and poles, but few saplings and seedlings. "There's nothing on the understory," Jim says. "It's just like a park." This woodlot needs to be thinned of the poor trees and the mature trees so that more regeneration will occur below them.

While Jim marks the trees to come down, I watch and his dog Skila wanders. At first glance, Jim looks clumsy. He moves about and stumbles, almost falling, several times. Soon I realize that sometimes he doesn't watch carefully where he's going. Instead, he cranes his head upwards, sizing up trees to determine their shape and condition. Walking around the tree, he always looks up and assesses it from all sides before deciding to leave it grow or cut it down. When debris on the forest floor lies in his way, he missteps a lot, tripping on brush and dead branches, so that it's a wonder he doesn't fall. Instead, he slips from one recovery to another, like a nimble toper. His sense of balance, I decide, is actually quite remarkable.

A tree in the forest is different from a tree in town. A tree in town, which gets lots of sunlight and grows many branches, is what Jim calls "limby." In contrast, a forest tree competes with other trees for sunlight, which means that it tends to grow straight and have no branches for thirty to forty or more feet above the moss beards that cover its base. Foresters call this long, branchless portion of a forest tree, not the trunk, but the bole or stem. The bole is the first thing Jim checks when he walks around a tree. He inspects it for any number of defects, including cankers, conks, rot, dead branches, forks, growth in clumps, broken-off limbs that leave scars, vertical seams running for many feet, and plain old crookedness—anything that downgrades or ruins the lumber. He also checks the canopy to see what the open space provided by cutting one tree will do for its neighbor.

Jim comes upon a maple with forty-eight feet of bole. It has a fungus and is almost dead; down it will come for pulp. He walks up to four trees between six and ten inches in diameter, but only four feet apart. They aren't big enough for saw logs, so he marks three for pulp to give the biggest room to grow. "Even though this isn't the best tree," he says, "it's the best I have right here." There are no absolutes in this work, he explains—no one ideal size or one gauge of quality. Each tree is judged with respect to its neighbors; a cull tree in one place might be a crop tree in another.

Jim is an old hand at telling the best from the worst. Because of his research at Michigan Tech, sawmill operators can now predict the quantity and quality of lumber that they should get out of their logs. In fact, Jim's work has resulted in the formulation of national standards to evaluate sawmill efficiency. That accomplishment was no overnight feat. It entailed cutting and milling two hundred thousand logs, and collecting and analyzing data on about fifty thousand of them. During the two decades it took to do that work, Jim came to appreciate northern Michigan winters. "We acquired so much information, you'd be lost if you didn't have the time to summarize and file and prepare maps," he says. "We needed time to get the records straight and do publishing and planning." That time was afforded by the winters when field work stopped and paperwork began, he says.

Jim's work translated into so much knowledge on yellow birch, hard maple, and elm, that he is considered a national expert on northern hardwood management in the U. S. I ask him about that reputation. He tells me modestly that his stature results primarily from circumstance. It's unusual, he says, for a forester to spend thirty years working on the same forest; a forester generally moves around more. His extended stint enabled him to undertake long-term research that other foresters haven't had the opportunity for.

Jim talks while walking and systematically sizing up trees. He marks a big oak about twenty inches in diameter. I notice that he paints it at eye level and also on the base of the tree, what will become the stump. That orange spot is a policing gesture. "I mark the butt to make sure the loggers don't cut an unmarked tree," he says, noting that, after logging, the owner can check to see if there are any stumps without orange markings, which in effect would amount to the cutter stealing trees.

Jim explains that there is always a debate in lumbering circles between the marker and the cutter. Loggers often want to cut more trees than markers designate; markers want to leave more trees for the future. Loggers, Jim says, "tend to not see the whole picture. Often they only see the immediate financial situation." Common practice in northern Michigan, he says, has been commercial clearcuts like Swede had to do—fell, say, every tree that is twelve inches or more in diameter. Everything left is essentially the same size and the same age and grows in thickly—what Jim calls "dog-haired reproduction." Such a practice is the exact opposite of achieving a stand with the distribution of ages—the four stories, he says.

A marker confronted with a same-size or even-age stand must open it up every ten years to let some trees grow big and let regeneration occur, Jim says. It takes thirty to fifty years to re-establish an all-size hardwood stand after it has been cut into a same-size stand, he adds, noting that logging companies sometimes maintain an all-age stand for years and then, just before they sell the land, commercially clearcut it to get as much money out of it as possible. "That's the trouble," Jim says. "In the end, economics often dictates the management—to the detriment of proper management." On northern Michigan private land, well-managed stands are rare, he says. Unmanaged or mismanaged land is the rule. "I go on different properties all the time, and you're still removing a lot of junk trees," he says. With the exception of having no saplings or seedlings, the woodlot he and I are on is a good timber stand, he says.

Jim calls Skila. The dog is ranging, but not far. He comes, wet and friendly. I am wet, too, from the rain-soaked brush and ferns.

When Jim sees a tree, he does not see shade cast at midday, leaves soughing in the wind, or any other aesthetics; he sees dollars. He evaluates the quality of the tree for pulp, lumber, or veneer and harvests accordingly. He looks for the highest dollar yield per acre. He constantly uses economic terms as an analogy to describe what he's doing. Of immature trees, he says, for instance, "Your bond hasn't matured because you still have a chance to gain quite a bit of interest on that tree."

The worth of a tree—and the gauges of its worth—vary. To figure the value of a potential harvest, Jim tallies in his notebook the species, diameter, length, and grade of every tree that he marks. The first criterion, species, is very important. In late 1994, for instance, hemlock, the most common northern softwood, was selling for a tenth of what plain maple was worth.

The second and third criteria, diameter and length, go hand in hand. The diameter of a tree is always measured at four feet, six inches above the ground, what foresters call "breast height." Saw logs must be eight to sixteen feet long. To have the minimum diameter of twelve inches on the small end, a sixteen-foot-long log must have a diameter at breast height of sixteen inches. Therefore, a tree that is sixteen inches at breast height is the smallest tree of saw log or veneer quality that Jim will mark for cutting. When Jim tells me this, I try to embrace a tree of that size. It is big; my fingers just touch. But Jim rarely marks a healthy sixteen-inch tree because that size isn't optimal; instead, he'll take out any competition and let it grow to eighteen or twenty inches in diameter so it will yield more than one saw log. To get two saw logs, a tree must be eighteen to twenty inches in diameter at breast height. To get three saw logs out of it, a tree must be even bigger. Jim says that good trees with three saw logs are rare on unmanaged private land in northern Michigan, such as the stand he is marking when I accompany him. Two-log trees are more common.

The last criterion that Jim tallies is grade. There are four grades of saw logs. The best, veneer quality, is a log that is virtually knotless and defect-free. The worst grade, wood that is used for crating and pallets, is full of knots and other defects.

After tallying these four factors—species, diameter, length, and grade—Jim knows what trees are to be cut in the stand, how many saw logs and veneer logs they will yield, and how much pulpwood they will yield. Then the owner knows roughly how much timber he has and what it's worth, and he is then ready to invite loggers to bid on the rights to cut it.

Jim walks on and bypasses a clump of oak—five trees of a foot or more in diameter growing out of the same spot. Oak, red maple, and basswood—in contrast to hard maple—grow well in clumps, he says, and he leaves them unmarked. "This is a heavier clump than normal," he notes, "but they're too small to cut and they have good form." He marks several deformed trees—he calls them "poor-formed"—for pulp and then cleans the nozzle of his spray gun with a maple leaf.

He continues on, barely pausing. He sprays a twenty-four-inch maple, a nice tree with three logs in the bole. He explains that the tree is mature, so it is good to cut it now and give other maple around it more sunlight. When a hard maple reaches the twenty-inch class, he says, it grows more slowly, eventually just maintaining itself, and after that deteriorating. In northern Michigan, he says, hard maple normally grows about a tenth of an inch in diameter a year—about the thickness of a nickel—and a foot in bole length a year. Oak, with an enormous crown compared to a maple, "grabs a lot of sun," Jim says, and grows twice as fast. Yet, in a maple stand, proper management can double the rate of growth on younger trees, to about one inch in five years, which is "excellent" growth, Jim says, for a sugar maple in the Keweenaw and a good reason to let younger trees supplant mature trees.

He marks a maple with a seam. A seam is caused by a frost crack. Wood expands and contracts, depending on changes

52

in moisture and heat. In the winter, wood contracts and the moisture in it freezes and expands. The resulting stresses sometimes cause a tree to split open. A seam, which can extend right into the heartwood, is a healed frost crack and often has ingrown bark over it. It is a serious defect in a log: the side with the seam will fall apart when cut into boards. The other three sides, though, will yield lumber—unless a seam spirals around a tree like a corkscrew, ruining the whole thing for lumber. Such a tree, even if it is big, can be used only for pulp.

Jim marks a crooked maple and another with a scar where a limb has broken off. As it grows, a tree, when young, will prune itself of small branches, and the knots disappear under subsequent sapwood growth. But a big limb scar won't grow over and will ruin that side of the tree for veneer.

Jim marks several trees and says, "You get areas like this where there are a lot of deficient trees, and you got to take the worst and leave the best of the worst." In the next area of the forty, he says, "It's overcrowded here." He marks a twelve-incher, one he normally would let mature for veneer. He marks a clump of maple, moves, comes across a good tree, and takes out a competitor that has three forks. He moves again and marks a tree that he judges to have poor potential. He first takes out hazardous trees, then defective trees, then mature good trees. "There are a lot of decisions you make in a day," he says.

We come to a poor area with many culls. Jim marks virtually every tree—ten of them in a fifty-foot circle. "We're in a bad spot," he says. "We don't have much to leave, but you've got to get some of these openings in here."

I look overhead and notice that the forest is deceiving. From the ground, as I look at the stems, the trees are twenty and thirty or more feet apart and seem to have plenty of elbow room. But looking up, I see that the crowns fill the sky and virtually block out the sun in places. That full canopy causes the park-like nature of this particular woodlot, Jim says, and opening up the canopy will bring sunlight to the forest floor and encourage the growth—the release—of the understory.

We come to a twenty-four inch maple. "Now, this is a nice tree," Jim says. "Beautiful tree. This is a money-maker for the cutter." He explains that loggers usually won't cut a timber stand unless it has at least a thousand board feet of timber and three or four cords of pulpwood per acre. It takes about fifteen logs between eight to sixteen feet long to yield one thousand board feet. Getting that amount of timber per acre in an unmanaged stand in northern Michigan is difficult, Jim says. But good stands managed by the Department of Natural Resources or the U. S. Forest Service yield two to three times that much timber, he says. Old maple stands on the Huron Mountains—the biggest hills in the U. P., located to the east and southeast of Keweenaw Bay—are simply spectacular, Jim says. He's seen maple stands in that country that haven't been logged for decades and would yield twenty-five thousand board feet per acre.

Jim sprays a nice oak. When he marks a tree, he girdles it with paint so the logger can see the orange mark from any direction, and he moves quickly and resolutely from tree to tree. As I ask questions of him, I keep moving in circles to avoid his aim. Once I'm not fast enough; Jim sprays me in the face, shirt, and pants from fifteen feet away. Apologetic and solicitous, he wipes my face with a handkerchief for several minutes. The paint won't come off my shirt and pants. It is the price I pay for getting caught flat-footed around a single-minded marker.

Jim calls Skila, who comes, wet and panting. In the summer she likes to roam in the woods. In the winter, Jim says, she won't venture far in the snow. She'll follow Jim's snowshoe tracks that pank the snow and then sit patiently where she can see him. She moves only to keep him in sight.

Jim comes across a sugar maple with a slight bulge in the bole, and his keen eye detects a slight lightening of the bark.

54

The bulge and whiteness are symptoms of a disease, one of those mysteries, he says, that foresters don't know the cause of. The tree gets patches of ingrown bark that ruin the wood for lumber. When working in the Alberta forest, Jim used to notice the bulge and the light bark. When milling the logs, he used to notice the ingrown bark in the lumber. Over time he correlated the two phenomena. "You can see how much experience is important in this kind of work," he says. "You don't get that in books." He then complains, "You can get a doctorate in forestry these days and not know how to spot white bark."

Jim comes across what looks like a nice maple tree. For the first time this day he sounds a tree—thumps it with the butt end of his small hatchet. "Rotten," he says, adding that the sound of hollowness is different in each tree. "Now, basswood always sounds hollow," he says, "but the more you do it, the more you can tell." When he was young and first started marking, he used an increment bore—a hollow, round, threaded device that screws into a tree and takes a plug out—to "tune" his ear. The increment bore will stop boring if a tree is hollow, and Jim would tap the tree with his hatchet, then bore into it to test his conclusion. One time he sounded a tree and a bear came out of a hollow on the other side of it. "We have excitements like that," he says. Shortly, he changes paint cans and buries the used one beneath a dead log. "By the time it's found," he says, "it'll be rusted to nothing." I notice that his fingers are stained with orange, like his carpenter's vest.

The mosquitoes this day are pesky. Jim talks about the black flies earlier in the summer. "You could never live out here without wearing a net," he says. "That net has a plastic top on it, and they're hitting it continuously. You think it's raining." So far this spring, he's found only four ticks; none bored into his skin. The ticks carry Lyme disease, which can be serious, even fatal, for victims, but they are merely a trade hazard for Jim.

We come to a huge, twenty-six inch maple. "There's no use leaving it here," he says. "You might as well start growing a new crop." He comes to a sixteen-inch tree that is tall, in fact, the tallest of the day. "Basswood," he explains, adding, "Now, there's a beautiful tree. It's an indicator of a good site. Very good, workable wood, and it grows limb-free like that. We won't mark that tree, it's not ready. It sure has nice potential. All veneer. If I cut that tree now, I'd get one veneer out of it because the upper logs would be less than twelve inches. If I let it grow to twenty inches, you'll get two or three veneer logs. I'll get compound interest. You've got to put your growth in good trees." With bad management, he adds, "you lose what you gain" by increasing growth on bad trees while cutting all the good ones. Jim tells me that at the Alberta center, an unmanaged forest stood across the road from the managed forest. Tourists were often asked to look at the two forests and to pick the more beautiful one. They invariably chose the managed forest, Jim says. Still, he notes, some environmentalists say that trees shouldn't be cut. "The average person doesn't realize the potential of these stands to continue growing after logging," he says.

He talks again about the park-like nature of the woodlot and the ease of moving through it. He says he also is marking a thousand-acre plot with a big, black spruce swamp. One day the week before, it took him an hour to walk a mile. "That's a long time to hike a mile," he says, "so you have those rough days along with these nice days."

I ask him how he keeps his bearings in such a big forest.

"That's one of the first things a forester learns," he says. He gives me a short lesson in measuring distances and areas. The key measurement is what foresters call a chain. A chain is sixty-six feet. Twenty chains is a quarter of a mile, the length of a forty-acre plot. Eighty chains, then, is the length of four forties end to end, or a mile. Eighty chains is the side of one section, which is a square mile and contains sixteen forties. Thirty-six sections is a township. A forester knows how long his step is and counts chains to know how far he's traveled. With his step to show distance, with a compass to show direction, with survey marks to show the corners of most forties, with land features such as rivers, gullies, and hills as guides, and with maps to correlate it all, an experienced forester can travel in the forest without losing his way or marking the wrong land. Sometimes foresters are aided by old blazes that mark the edges of forties. If those aren't evident, all Jim does on a big plot is find the corners of the land, pace off and paint the edge of the forties, and then, using his paint line as a guide, mark the rest of the timber. That's the ideal. But sometimes the forties don't have any survey marks, and finding the corners takes, well, "It takes some effort," Jim says. This day's small plot outside Chassell has survey marks on the corners, so Jim may as well be in his backyard.

In some small sections of this woodlot, tiny sugar maple seedlings carpet the ground. They are thick, as alike as drops of dew, and they cover the ground like grass. They will bide their time in the shade and eventually grow to become the dominant species. But maples have yet to take over here, and midway through this stand, Jim has marked 154 trees for saw logs: twenty-three hard maple, forty red maple, and ninety-one oak. Jim points to the seedlings on the forest floor. "Imagine what it's going to do when I get these trees out of here and get some sunlight in," he says. "This site will convert more to maple when we selectively log it like this. The maple will increase tremendously." I ask him about damaging the seedlings by stepping on them. "There's thousands more than you need," he explains.

We walk through the seedlings into a small, shallow drainage.

"Drop it," Jim commands as we begin to walk up the other side.

I think he's talking to me and am puzzled. Then he points out that Skila, quick as a snake, has snatched a small bird from the branch of a seedling.

The dog obeys.

"See how he dropped it," Jim says. "I made him drop a chipmunk last week. I try to train him not to be a killer, but there's that instinct in them. He came right up on it and grabbed it."

He picks up the bird and examines it.

"I'll leave it," he says. "Nothing we can do."

He places it on the forest floor among some seedlings. A couple of them tremble where the bird lies twitching. We continue with business.

Jim says that marking timber is more of an art than a science, but that it yields very practical results. "Right now we're trying to give the better trees room to grow," he says. "The next marker's got to make decisions, too. You can make a second judgment on it in the future. It'll last."

We come upon an eighteen-inch hard maple. "That's a dandy," Jim exclaims. "Too good to cut now. All veneer. That's what you've got to leave if you're going to manage for the future. A lot of loggers will look to take that kind of tree. Say, 'To hell with the future—give me my dollars now.'"

We skirt a farmer's field. "Oh, oh!" he says excitedly. "Look what we're coming up to. Our biggest tree yet. Wow-whee! It's a monster! That's oak for you. She'll grow big." It is thirty-four inches and overlooks the field, which means it has had lots of sunlight and is "limby," so there is no veneer. "But," Jim says, "they should get some good lumber out of it." He sprays it.

We come to a beautiful, big hard maple. Jim sounds it, tiptoeing to reach high. "As far as I can reach, it's hollow," he says. "Oh, and it's bird's-eye." He points to the bark and says, "The bark is dimpled. It's not always that obvious." He knocks a bark chip off with his hatchet. The white wood beneath the bark is wet, as slippery to the touch as fresh orange seeds, and, like the bark, it, too, is dimpled with little bumps. But a hollow bird's-eye is like an empty promise. I feel a pang of regret for the owner, but Jim, all business, just paints the tree and moves on.

I ask him if he enjoys his work.

"People go to the woods for recreation," he says. "I have my recreation in my work. You're not so ready to go wandering in the woods after a day spent marking timber. You wouldn't want to do it day in and day out. You're glad to see it's finished. Still, when a job is done, you come back and you're all renewed again."

The only other time he roams the forest is when he hunts deer. "When you're hunting, do you look at a tree as something different than a log to be sold?" I ask, thinking figuratively.

"When I'm hunting," he says, thinking literally, "I don't look at trees. I look for deer."

We cruise for another hour. Jim has unwavering, intense concentration for the task. At noon he quits for the day. I thank him for taking me along.

"Did I distract you?" I ask.

"No, not much," he says. "It's good to talk to somebody that can respond. The dog comes up and just licks you."

John Porkka

THE LOGGER

The timber marker is the benevolent despot of the bush. His word goes, and, ideally, he ensures that the forest will grow tall, straight, and true and that the logger will have a job today and the promise of a livelihood tomorrow.

John Porkka relies on the good judgment of people like marker Jim Johnson. A jobber for Mead Corporation, John, who was born in 1959, has logged in the Keeweenaw since 1986. I was told that he is accommodating, and on a mid-June day I call to ask if I can spend time with him on the job. He tells me to meet him at half past four the next morning at his home. I roust myself out of bed at half past three and make my way to Liminga, which is not a town, but an area of woodlots and potato farms a few miles west of Houghton, near the little settlement of Atlantic Mine. In the darkness I don't know which driveway is John's, so I pull off the road and open my window to listen for activity. The morning is cool, and the air is heavy with the moisture of last night's rainfall. Strings of light fog rise from the wet roadway. Fireflies float lazily about and look like shooting stars. I hear John's truck idling, and find the long dirt driveway that takes me to where he is warming up his truck. I introduce myself, for we have only spoken on the phone. He is nondescript in the semi-darkness on the edge of the truck's headlight beams. We climb into the cab and begin what will be a long, wearisome day.

John runs John Porkka Logging, which consists of three pieces of equipment—a tractor-trailer with a loading boom, a bulldozer, and a skidder. He employs three men besides himself—two sawyers and a man to skid logs to clearings, called landings, along the bush roads, where we're headed. The first thing John does is apologize for his truck. "It's an older truck," he says, "but they cost a few bucks still." He bought it six months previously for fifty-one thousand dollars. It is nine years old and has three hundred thousand miles on it. "As long as the cab and chassis stay together," he says, "you can keep them going forever." John worked on this truck until nearly midnight the night before. The throttle line needed replacing, and the mud flaps had to be sized down and remounted after he tore them off by backing into a dirt bank the day before. He has had four hours of sleep.

John is logging land near Painesdale, the northern edge of the country that Fred Aho liked so much. The trees that John's crew cuts are on Mead land and are marked by Mead foresters. "They have it marked so it's not a real heavy cut," he says. "We're covering a lot of ground to get what we do. They mark it so the jobber don't go in and take all the good ones, so it isn't overcut. They're doing a pretty good job— where a guy can make a living, too, you know. They take a fairly good cut—not too heavy, not too little. It's good for the jobber."

This jobber has found his calling. "I wouldn't do nothing else unless I had to," he says. His contentment contrasts to the restlessness of his young manhood, when he bounced around from job to job. "All the other jobs I had, I lasted about a year, ten months," he says. "Get tired of it and go to something else. This is enjoyable for me. Probably not for a lot of people. But I like it." At first he logged just to keep money in his pocket. Now the tiring routine wears well. "You can work as hard as you want and as long as you want," he says. "You can keep production. I enjoy it, probably, is the biggest thing."

A potato farmer by upbringing, a logger by choice, John is mindful of the weather. Later in the day he will say, "When you farm, you want rain. When you log, you want it dry." Now, in the pre-dawn, the damp air prompts him to talk about the wet spring that made it difficult for him and his crew. The land that he's logging has some clay in the soil;

when it's wet, the road is slippery and he can't get his truck into the bush to haul out the logs that his crew has cut. That won't be a problem today. About two miles southwest of Painesdale, John turns off the wet blacktop road onto a narrow dirt road, and right away he says, "It rained just enough to keep the dust down. It was really dusty yesterday. Couldn't even see behind you." A minute later he adds, "The mosquitoes will be out in full force this morning with that moisture."

We lumber along the bumpy bush road, and the headlight beams jerkily pierce the darkness. After ten minutes of slow progress, it is 5:20 a.m., and we are at Eleven Mile Lake. It is only ten acres in size, but is a big enough opening in the forest for us to see daybreak on the horizon. John has to back up the rig, which is about six times longer than the average car, for a quarter of a mile. "This is easier than backing up a hay wagon with a tractor," he says. Soon we are at the landing where the logs are piled, and it is light enough for me to see John. He wears dirty jeans, a blue and yellow plaid shirt, leather boots, a maroon baseball cap, and dirty gloves. He has blond hair and wears glasses. Of medium height, he is lithe and wiry, and like water pouring out of a bucket, he slips fluidly and gracefully out of the truck's cab and to the ground.

John's tractor and tractor bed (he calls it the "rack") is about forty feet long. The trailer that hitches to the tractor bed is about twenty-five feet long. Situated on the back of the truck bed are the controls of a loading boom that is twenty-seven feet long and hinged in the middle (hence its nickname: "knuckle boom"). On its far end, the boom has pincer-like claws, called a "clam," for grappling logs. At rest, the boom lies about ten feet high over the truck bed, and the clam rests on a steel deck above the cab. At work, the boom pivots in a full circle to load or unload both the truck bed and the trailer.

With a nimbleness that only the young know, John climbs aloft, fifteen feet off the ground, to the seat of the hydraulic controls for the boom. He immediately extends big, flat, steel feet that lift the truck off of its rear wheels and stabilize it. Then, with the truck engine idling fast to provide power, John puts the apparatus to work. The claw that hangs from the boom is like fork and spoon: one half is two curved prongs; the other half is a curved metal plate. They work like jaws and grasp from one to five logs at a time. All of the pulp sticks, which are headed for a paper mill, are as long as the truck and trailer beds are wide: eight feet. John loads the sticks crossways. After each clam-full is situated, he swings the boom and claw sideways and gives the sticks a little bump to align them evenly. The boom and claw are like extensions of his arms and fingers. Later he will say, "It takes a lot of coordination to learn what lever to use. Now you don't even think about it. It's like playing the piano probably."

John approaches the work with the persistence of a mosquito. Mosquitoes are on my mind, for John's prediction is accurate: they are in the air as thick as rain. I sense an itch on my finger, look, and there's a whole family on one knuckle. Trying to chase them all away, my arms are like windmills on a blustery day. Meanwhile, John works five hand-levers and two foot-pedals, and his concentration is broken only by an occasional spit of chewing tobacco. Not once in two hours of loading does he swat a mosquito. "Oh, you get used to 'em," he says when he climbs down from his perch. It is only eight o'clock, and he has already been up and working for four hours. He will go steady for more than another ten hours, taking only a few minutes for himself. He appears indefatigable.

The pulp sticks that John loads are from four to thirty inches in diameter, and they are all hardwood: hard maple, oak, white birch, and a little bit of basswood. (Only 5 percent of the load can be basswood because too much of its bark fouls up the paper-making process.) John hauls the pulp to Champion International Corporation's mill in Quinnesec, Michigan, 150 miles south of where he is cutting, and to James River Corporation's mill in Amasa, Michigan, ninety miles south. Both mills contract with Mead for pulpwood. John's crew cuts both saw logs and pulp; about 70 percent of the cut yields saw logs, mostly hard maple. In contrast to his long pulp runs, John has only a fifteen-mile trip to deliver saw logs to Mead's Northern Hardwoods Division, which is halfway between Houghton and South Range.

After loading about 250 pulp sticks, John climbs down from his boom seat and secures the load with two chains. Then he gets into the truck cab and in low gear creeps out of the bush. He says the truck is little different from an over-the-road rig. The only added feature is a differential lock that allows him to use both of his tractor's tandem axles for pulling on bush roads. When on a highway, he switches the differential to one axle. We reach the highway about half past eight. John drives a few miles, then stops and gets out to check the chains. With the bouncing and jostling caused by bumps in the road, the load often shifts and the chains loosen. When he gets back into the cab, John says, "Yah, they were loose. They usually are. When you hit them bumps, it settles in there."

He opens up a can of Squirt, drinks some, puts the can on the floor, and lights up his first cigarette of the day. Then, in a voice that is oddly deep for so lean a man, he talks about his business. He's careful not to overload, for the fines are big. Still, he's hauling nearly 100,000 pounds of logs to Quinnesec. That load, plus the weight of his rig, means that more than 150,000 pounds—seventy-five tons—are rolling down the highway. That's a lot of weight to get up to speed, and John drives ten miles before the terrain and momentum allow him to shift into high gear. "One thing," he says, "you know where the hills are when you get in one of these, even the little ones." His truck is equipped with a Jacobs Engine Brake—what he calls a "Jake Brake"—which, when switched on, chokes off the fuel intake but maintains compression for slowing down. It saves on the brakes. It kicks in automatically and noisily when he decelerates.

John doesn't like driving the truck. "You really get nervous," he says. "I like dumping trees—felling. That's the best work. I don't mind doing this, but I'd rather be in the woods. But what are you going to do? You gotta move your wood." A little while later he adds, "You know, a tree—you can tell what it's going to do a little bit, anyway. On the road, you don't know if somebody's going to pull out in front of you. I'm getting better. I'm more relaxed. When I first started, I'd get pretty tense. It's tough to get used to."

When John doesn't haul saw logs or pulp sticks, he maintains equipment or fills in when one of his crew is off. He knows how to do each job that his men perform, an ability that he attributes to growing up as a farm boy. With all the work that farming involves, he says, "You see what's going on. You learn a lot when you're young."

John has just started to work for Mead. Before that, he worked for another big company that operates in the Keweenaw. He likes Mead better. "It's a good outfit to work for," he says. "Better timber, they put some money on their roads, and the pay is better." John says his brother Bruce also logs and deals with private landowners, bidding on their timber and hassling with road easements. "I'd just as soon work for a company," John says. "Get up, get it done, no headaches."

The only thing he dislikes about his work is the paperwork. "I'm not one for it," he tries to explain. On the dash of his truck hangs a big magnetic clip that clasps a thick bunch of gas receipts, scale slips, maintenance records, driving and insurance papers, payroll information, and more. "You can see I'm behind," he says. He hires somebody to do most of it. "If I had to do all of the paperwork, I'd probably quit. That's the worst part about this job." Such duties have always bothered him. He is meant to work with his hands.

66

When he was in school, he was uninterested and inattentive. "I never cared for school," he says. "Fought it all the way through. I went, but it's nothing I enjoyed to do. I took my books home just to make it look good, I guess."

After he graduated from high school, he took a one-year course in mechanics, farmed for a year with his father, worked as a mechanic for a year in Dollar Bay, then went to Texas and worked construction. He did not like the city, and to illustrate his distaste he quotes one of his brothers, who told him of city life, "'You look out the window, and it's better than watching TV. Shootings and sirens all the time.'" John says awhile later, "My ma and them wanted me to go to college. I always told them, 'Somebody gotta be the ditch digger.'" After a year in Texas, John moved back to the farm to help his father. Ever since he was young, John cut firewood on the farm for "extra money for the potatoes." Back home again, he continued to do so. "One thing led to another," and he began logging.

As he talks, John keeps both hands on the truck's big steering wheel. He waves at all passing truckers. He takes his second chew for the day, spitting into a styrofoam cup that he places in a big, handleless coffee mug on the floor. We're forty miles down the road now, and John stops one more time to check the chains. Back at the wheel, he says, "They were pretty tight, but I tightened 'em real good. They'll be okay now."

John didn't have the money to get into logging. When he started, he purchased a bulldozer on a lease-purchase arranged by the manufacturer. About 80 percent of his rental payment went toward a down payment, and three months later he bought the machine at 12 percent interest. "I didn't have any bankroll," he says. "In fact, that first year I had to borrow money for Christmas." When he bought his skidder a few years later, he again financed it through the manufacturer, this time at 13 percent. When he bought the truck, he had saved up $10,000 for the down payment and was able to finance the remaining $41,000 through a bank, again at 13 percent. He has paid off the bulldozer, which cost $16,000, and the skidder, which cost $21,000. His truck payment is $1,400 a month.

These expenses are partially offset by John's pulp runs. He gets about twenty dollars a ton for taking a load of pulp to Quinnesec. That rate works out to about a thousand dollars a truckload (about four bucks a pulp stick). He figures he clears about two hundred dollars. He hauls four loads a month to Quinnesec and the same number to Amasa.

John says that the previous year he grossed $175,000, paid $130,000 on equipment, labor, insurance, taxes ("They like to get them taxes. I paid a man's wages in taxes"), and other overhead, and took home $45,000. His contribution to workmen's compensation is about twenty-five cents for each dollar his crew earns, the highest rate in Michigan, he says, due to the dangers of logging. Nevertheless, John expects to make more money this year because he has his own truck and doesn't have to contract for hauling. "It's a good tax write-off, the truck, because there's so much expenses, plus you own something in the end," he says. Unlike others he knows, he doesn't want to get much bigger. "A lot of crews go that way, but when the market gets bad, they're going to feel it," he says. "They still have to make the payment. There's not too much guys can pay cash." Later in our journey, he adds, "I don't know how some of them guys figure it. They go all out and buy everything. I don't know how they can do it. . . . I don't figure to get that way. The way I am I feel comfortable."

Nevertheless, a major worry and expense for John is the maintenance on all his equipment. In the last two years alone, he has rebuilt the engine on his skidder ($4,000), rebuilt the transmission on the skidder ($5,000), and fixed the brakes on his trailer ($2,500). Even everyday things come with princely price tags. A tire for the truck costs $300. An oil change costs $80. Fuel for the run to Quinnesec is almost

$80. Soon he will buy four more ten-foot-high steel stakes for his tractor bed ($75 each). On top of all that nickel-and-dime stuff, "free labor you don't count," he says. Trucks and skidders and bulldozers, he says, "they're a lot of work, that's the truth." I tell John that I know of one Keweenaw logger who still skids with horses. John replies, "You wonder how they did it years ago with horses. You got to get along with those things, too, you know. Bring them hay and water. It ain't the wave of the future, that's for sure."

Thinking about all the machinery and the responsibility involved with his operation, I tell John, "It takes guts to do what you do."

"I know it does," he answers. "I feel that, too. There must be pressure, I guess. I had a problem here a year ago—never felt good, always tired. The doctor said, 'Oh, you got too much stress.' I said, 'Well, everything's going good.' But then I realized later that, you know, I'm worrying a lot."

We approach Quinnesec. We have come about 130 miles, and the roads that we've been on have parted the bush only tentatively. We've passed through few towns. In fact, from Copper Harbor, the northernmost town in the Upper Peninsula, to Quinnesec, which is near the Wisconsin border, trees hug the road. "It's a lot of wood," John says.

We reach the Champion mill. I have to wait in the security building while John goes into the yard to unload. He says that he'll be back quickly, that it takes less time to unload than to sweep the bark and dirt off the trailer. He is gone forty-five minutes, though. During that time, sixteen other pulp-laden trucks pass through the gate; an equal number come out. Each truck is weighed with its load, then weighed again empty, and the difference, plus mileage, is the basis for payment. John's load weighed just shy of a hundred thousand pounds. The Champion mill, he says, buys about 120 truckloads of pulp a day.

As we pull out of the mill site a little after noon, the truck is noticeably peppier and accelerates more quickly. John dips into his lunch and eats as he drives. He cracks open another Squirt and has two sandwiches, a banana, and yogurt. After he eats, he lights his second cigarette of the day.

We soon pass a potato field. It is still raining. "The potatoes are going to grow now, with a little water on 'em," he says. "In a few days, you'll see those plants get big." He wonders whether it's raining up north, near Painesdale, and whether he'll be able to get into the bush and load up with saw logs for Northern Hardwoods. "This kind of weather is best for sleeping," he says. "Nothing else you can do. You don't feel guilty."

A short while later we are in Crystal Falls, about thirty miles west of Quinnesec. John stops to buy a cup of coffee to go. Five minutes later, we are back in the truck and on the move.

Sleep apparently is often on his mind. He said on the way down, "You don't want to do this every day. You get tired just from the driving." Now he explains that he gets up at four o'clock only when he makes the pulp runs. Usually he begins work at half past five or six o'clock. Lately, he's seen a lot of nineteen-hour days. He is weary but likes the variety in his work. The previous week, one of his men was off, so he skidded logs. "I like it out there," he says. "It's nice, you know. You don't get so tense like driving this truck." He learned to drive by going to truck-driving school, then by just doing it. He's learned to drive all his heavy equipment that way. "Kind of self-experience, I guess," is his way of describing it. Although driving the truck wears on his nerves, he says that he's not bored with logging the way he always was with his other jobs.

"Why were you bored?" I ask.

"I don't know," he answers. "Do the same thing every day. And a lot of people get to you, too, you know. Out in the woods, you don't argue with anybody but yourself. Something goes wrong, it might be your own fault."

He says that his skidder operator is a mechanic who also prefers to be in the bush. "I think that's why most people are out in the woods," he says. "They enjoy being out there."

"I couldn't stand the bugs," I say.

"Well, you got to live with 'em," he says. "They get bad, I guess. At least they go away after awhile. When you have them other jobs, some problems don't go away until you quit."

Workers in the northern bush are pestered by bugs. The first year John logged, the black flies were particularly bad. "Your whole chest would be black with 'em," he says. This year they were so mild that for a minute he can't remember if the black fly season was over or was yet to come. He dislikes deer flies the most. "Them deer flies are the worst for biting," he says. "They take off with a chunk of your skin when they go."

John is happy with his crew of three. He has one felling sawyer, who drops the trees, the skidder operator, who drags the long logs to the landings, and the landing sawyer, who cuts the logs to length. The felling sawyer is paid by the board foot and can make about twenty-five thousand dollars a year. The other two jobs pay about thirteen or fourteen thousand dollars a year. The felling sawyer makes the most because, John says, "that's the most dangerous job, and he's probably working the hardest." He recalls one summer a few years back when three lumberjacks in the Keweenaw were killed. "We haven't had any major injuries on our job yet," he says, "which ain't too bad considering woods work." The danger makes him a boss who is indulgent, particularly with his felling sawyer. "If it's real windy, he probably won't work, and I won't make him either. I don't blame him. It's dangerous with an off-and-on wind. I've had leaning trees pull back and pinch the saw. A steady wind is not as good as no wind, but better than off-and-on, gusty."

John's felling sawyer is only nineteen years old but has all the makings of a good laborer, because, John says, he makes every move count. "I've had other guys rush, hurry, end up getting less done in a day," he says. "This one's pretty good for a younger kid. He knows what to do. Some guys don't push too hard, you know. This young fella's got some vinegar in him." He'll drop about a hundred trees a day—counting the ones both for saw logs and pulp sticks. Cut by the landing sawyer into lengths from eight to sixteen feet, they make enough saw logs to fill the truck twice a day. John says lumberjacks tend to bounce around but he hopes to keep this crew together. "You figure to keep your good guys," he says.

The crew works six days a week. Mead plows the bush roads in the winter, so the only time the crew gets off, besides Sundays, is in the spring when the bush is too muddy and when the county road commission puts road restrictions into effect, which keep the heavy trucks off the secondary roads until the moisture from the thaw dries.

We arrive back at John's logging site again at half past two. John's men, who started near dawn, are knocking off for the day, and they come out of the bush just as we turn into it. John stops, grabs paychecks that he forgot to give the crew in the morning, gets out of the cab, and talks to his skidder for five minutes about the day's work. As he does, he takes his hat off and wipes a hand over his brow and through his hair. He stretches, bends backward, rubs the small of his back. Then he puts his hands in his back pockets. Along with the few minutes it took him to get the coffee in Crystal Falls, it is the second time all day he's stopped working. Shortly, he climbs back into the truck.

"They got a hundred-forty trees today," he reports. "Smaller ones, though. They were cutting roadside trees."

He explains that the foresters mark roadside trees, even if they're small, to let the sun in so that, when it rains, the road dries faster. I ask John whether he has ever got the truck stuck on wet bush roads. He says yes, several times, and that last week he had to use both the skidder and the bulldozer to get free. "Never had to unload it yet," he says.

We drive slowly toward the cutting site for the second time this day. Now that it is light, we can see the deeply rutted paths that the skidder left two weeks before when the woods was sopping wet. Tops of trees with dried or droopy leaves lie all along the edge of the road. The blond color of the stumps stands out amid the dark trunks and greenery.

We come to a fork in the road. John gets out and walks a few hundred feet to one site where the crew was cutting. The road is muddy. "Don't pay to monkey here," he says. He just wants to see how much has been cut. After he inspects that staging area, we take the other fork to a second cutting site. There is a low, wet spot just before this landing, too. "I won't be able to pull out of there without fixing the road up, and then it'll be real late," he says. He decides to unhook the trailer, leave it at the fork, and load only the truck bed.

After unhitching the trailer, John backs the truck through the soft spot and to the landing. He gets out and puts seventy-five-pound steel stakes on each side of the truck bed, for he will load saw logs now, and they lie lengthwise, not crossways, and therefore need stakes on the sides of the bed to keep them in place. Then he climbs to the boom chair, lowers the stabilizing feet, and begins sorting through the logs. He puts sixteen-foot logs along the sides of the truck bed, resting against the stakes. He puts shorter logs in between these. As he sorts through the pile of logs, he picks out pulp sticks and stacks them on the ground for easy loading in the morning. When he lifts logs, he turns them around in the air to look at both ends; logs with bad hearts or lots of knots are dropped on the pulp pile. Whatever he lifts—pulp sticks or saw logs—dirt, bark, and sawdust from the landing sawyer's cutting fall to the ground like hail.

It takes an hour to load the truck. It is another hour for me of battling mosquitoes. Once again John is oblivious to them. When he is finished, he enlists my help to chain the load. Then he gets into the cab and says, "Hope we get through that bad spot." He eases the truck forward. When we hit the wet area, we can feel the truck hesitate. "She ain't going to make her," John says, and then the back wheels settle in and the truck stops.

John hops out of the cab and walks down the road to get the skidder. It has big tires, is geared low, and is powerful. It pivots—what John calls "articulates"—in the middle—both up and down and side to side—so it is more flexible than a wrist and can turn on the proverbial dime. He shows me how to shift it and steer it. I try to pull him out, but the truck won't budge. He decides to pull and also use the skidder's winch, so we change duties—he runs the skidder and winch, and I engage the truck's clutch and steer. When he turns on the winch, the wheels of the skidder churn and dig, but, instead of the skidder pulling the truck out, the dead weight of the truck pulls the skidder backward.

John decides to pull the truck out from the rear. The road is blocked by the truck, so he skirts it by going through the bush. The skidder crawls over brush piles and dirt banks and pushes down saplings like a tank. He hooks a cable to the back of the truck. With me in the cab engaging the clutch and steering, he manages to pull it free. Then he walks down the road to get his bulldozer, returns, and scrapes away the mud from the low spot, which he then fills with dry dirt. This time John hooks up the truck to the bulldozer, and he rolls through the bad spot nicely. It has taken an hour to get free. After parking the bulldozer, John climbs to the cab and says: "No sweat. A challenge, anyway. As they say, that's logging."

When we get back to the fork, John hitches the trailer back on so he'll be ready to load first thing in the morning. Then, lighting his third cigarette of the day, he pulls away. A minute later he stops.

"Did I shut the skidder off?" he asks.

"I don't know," I say.

He stops, rolls down his window, shuts the truck engine off. It is the first time since before dawn that the truck has stopped running. John listens for the sound of the noisy skidder.

"That's one thing about me," he says. "I tend to forget. Let me think. Yah, I remember now. I shut it off."

He restarts the truck. As we ease along the bush road, he eats two raspberry crisps. "I don't gain any weight," he says. "I can eat and eat and eat."

After unloading, he hopes for an uneventful ride to his father's house in Liminga, where he will park his truck for the night. If nothing happens to the truck that will require repair, he will be able to go to his own home, eat dinner, and maybe take his wife for a ride to town. His wife will like that; she is alone a lot with two kids, and she is pregnant with another. "She calls this truck my second wife," John says. About not seeing much of his family, he adds, "I think you got to do that in logging if you want to progress at all. As the years go on, it'll probably slow down. It wasn't this bad until I bought the truck. With the truck, now, it's a long day. But I more or less had to do it. 'A necessary evil,' like they all say."

We drive to Northern Hardwoods Division, and John unloads fifty-three logs. It takes him twenty-five minutes. Once he mishandles the controls and drops some logs. John lays the logs in single rows on both sides of the truck. They can't be stacked until the scaler determines how many there are, how big they are, and what grade they are. When the last log is off the truck, John straightens the boom and rests the clam atop the steel plate over the truck cab. Then he climbs down from the boom seat and comments on himself, "I played dropsy with the logs. I dropped three of them. Must be getting tired." Then he marks his code on the end logs of each row so that he'll get credit for them. He figures the fifty-three logs constitute about three thousand board feet and that he'll get about three hundred dollars for them—much less than the market value because he's cutting for Mead on Mead land marked by Mead foresters. After he puts his stamp on the logs, John sweeps the truck bed off, then climbs into the cab. "That's that," he says. He takes his cap off and wipes his brow. He yawns and sweeps a hand through his blond hair. It is nearly seven o'clock, and he has been on the go for fifteen hours.

He drives toward his father's farm, about ten miles distant. On the way, I ask him about his attitude toward the woods and nature in general. It is the first time all day that he searches for words to express himself: "I don't think about it," he finally says. "I don't go overboard, like some of these ecologists—you know, 'Don't cut a tree.' I don't go and trash the land, but you got to cut trees to make a better forest. I never thought of that question. I'm not a tree hugger."

As we near his father's farm, he remarks how another logger's truck has broken down and the man has asked John to haul pulp to Quinnesec. "Well, I suppose I could haul a few to help him out," John says. That will mean another early day. That prospect reminds me of something John has said earlier in the day on the way back from Quinnesec. "You know," he told me, "a lot of people tell you, 'It's nice to be your own boss, eh? A day off when you want.' I think it's the other way. You tend to work harder."

Henry Clouthier

THE SAWMILL MAN

Swede Intermill likes to listen to logs being milled; he hears
music. Henry Clouthier likes to watch logs being milled; he
sees money. Swede hears a song, Henry sees a sale. Swede
hears a chord; Henry sees a cord. Old, mellow Swede
wouldn't get a word in edgewise with old, spirited Henry.

Henry is big and brawny, with arms like tree limbs and
hands like burls. His oversized body is matched by his oversized imagination. Together, they give him the energy of a
man much younger. Throughout his life, they also gave him
the wherewithal to succeed in business. "Whatever I tried, I
made money," he says. "I made money like water, because I
tried and I tried and I tried. I was never a guy, I mean, for
being disgusted. Don't know the word 'disgusted.' I say if
you don't make it this way, try that way. I keep going 'til you
got it. I run seven businesses. I retired at fifty-eight."

Henry's businesses involved trees. In fact, he has nearly a
lifelong association with hard maple. He chopped down
small hard maple saplings for his parents when he was
twelve, chopped down big hard maple trees when he was a
few years older, loaded high-quality hard maple boards on
ships, logged hard maple during the 1930s, 1940s, and
1950s; and in those same years he sawed it, usually into railroad ties, at several sawmills he built and ran. Henry says he
worked at twenty-five different trades in his life. Most of
them involved what he calls "bulling," by which he means
hard manual labor. "That's all we ever knew," he says. "Oh,
yah, you got to work. You don't work, you don't live long.
The body is made for work." Much physical exertion has
made Henry quite hefty, and his imagination is as imposing
as his big body. He tells me story after story about his working life:

About the baths that lumberjacks staying at his parents' boarding house used to take, two at a time, in a big tub in the open shed in the back yard: "When they drained the tub after every bath, the water was so dirty you'd think it was molasses."

About the drunk who once grabbed him by the throat and punctured his windpipe, what Henry calls his *collette*: "That sumbitch had a hand three times the size of mine, and that thumb was about that wide"—he spreads his own big finger and thumb two inches. The man's thumbnail was so long and hard and sharp, Henry says, "I think you could cut wood with it."

About men walking in circles in the bush because their judgment goes haywire and nature's one short leg crosses them up: "The mind and that short leg of yours is always bad."

About superstitious lumberjacks crushing the saints of heaven with their feet: "They were mad and lost their temper, and every time when they landed into some hard luck, they'd put the blame on God. And God had all the saints, so they called all the saints down and put them in their hat and jumped on it. Imagine that!"

About a rival lumberman whom Henry didn't like: "I don't think he could make a plug for his ass if he had a turd for a pattern."

Henry's imagination is as bold as a hungry fox, as gaudy as the autumn northwoods, as flighty as gossip. Some of his stories will confound the literal-minded person, for as Henry says, "If you're too honest, it don't work. I was very honest at one time because I was brought up that way, but through life you learn a lot of funny things, see. You gotta brag in this world. If you don't brag a lot, you don't get nowhere."

Henry was born in 1906 in Lake Linden. The town is inhabited by many French people, and in 1889 it even had its own French newspaper. Lake Linden's Catholic Church still bears the name *Église Saint-Joseph*, and the phone book today

80

is filled with names like Archambeau, Clairmont, Perrault, and Sanregret. Henry's father's original name was Cloutier; it was anglicized to Clouthier. The French pronunciation was *Clue-tee-ay*; it's now pronounced *Clueth-year*. Henry's mother was born in French Canada and his father in Mississippi. His father came to Lake Linden via Canada in 1872 at the age of fifteen. He was accompanied by Louis Goodreau, Henry's great uncle, who was a resourceful fellow: when he was in his 80s, he went on a drunk one winter night, passed out on the side of the road, froze part of his foot, and the next day in the kitchen cut off his big toe with a mallet and chisel to save the foot.

Henry was the tenth of twelve kids. "The more they came, the better they looked," he says. His oldest brother was eighteen years older than Henry, who saw hard times when he was growing up. "You had a big family—they ate," he says. "I never ate so good at home. I seen more mealtimes than I seen food." Henry himself had six children, five of them boys. Today he is the patriarch of five generations of the Clouthier extended family. He lives just outside Lake Linden, along the Trap Rock River, which runs through a valley by the same name.

Henry remembers the log drives on the Trap Rock early in the century. Lumberjacks used to dam the river in several places, let the water build up, and dynamite a dam to release the water in a rush that would send the logs to the next dam downstream, which in its turn would be blown open. This staccato journey down the Trap Rock continued until the logs were down to the mouth of the river at Torch Lake. During those drives, some of the logs would hang up on high ground as water receded from the valley to the river channel, and Henry enjoyed watching oxen drag them back to the river. "Oxens was stronger than horses," he says, "because they have that split hoof." As the horizon beckons the wanderer, one idea always leads Henry in unexpected directions, and the split hoof of the oxen causes him to add, "The dog is the strongest animal in the world according to weight because he has those toes. Sure, the dog is the strongest animal there is. You know which is the most vicious animal on earth?"

"A weasel?" I guess.

"No, the human being," Henry says. "You know why? When he's aroused, watch yourself. He stops at nothing, see. Why? He has a sense of reasoning with his madness. That's what makes him dangerous."

When talking about human nature like that, Henry is a curious blend of racial attitudes. He himself is part Indian on his mother's side, but he says that when he was young, Indians were a suspect lot. "They'd steal the Lord's supper and come back for the tablecloth," he says. On the other hand, he admires the spunk in what he calls half-breeds and says that his mother sometimes degraded, not Indians, but white men. She told him, "'The white man is the most hungriest son of a bitch I ever seen.' She must have learned that from the Indian. She said, 'He'd kill his mother for a bag of gold.' My mother was a tough egg, you know. Oh, yah. She wasn't very goddamn bashful."

The first time I talk to Henry, one of eight visits that stretched over three months, he coughs repeatedly—a loud, rasping hack. His voice sounds hoarse, and, at the end of a long sentence, it sometimes plays out to a squeaky whisper. He blames the problem on five pills that he's taking on doctor's orders for an enlarged heart, high blood pressure, and a cold. "See, my throat is aching when I talk," he says. "Those sons a bitchin' pills done that. I never had a pain in my throat. Now it's burning and cracking in here all over." His big fingers rub his neck up and down. "Doctors are skunks," he says. "My doctor said I had a cold. I told him it was the Lake Superior catarrh. He laughed: 'I never heard of that, and I been here for twenty years.' Well, cripes, I been here for many generations. All the French call it the Lake Superior catarrh. I said pret'near every Frenchman has that. I said they don't die from it, but it's bothersome. You can't cure it. 'I can cure anything,' he said. And that's what he's trying to do. He might kill me by doing it."

I often can pick out an old Frenchman in a crowd because he talks with his hands as much as his mouth. By that standard, Henry is French blue blood. His thick arms and big hands punch the air for emphasis. As he talks, Henry often sits on the edge of his chair and waits for the emotion of a story to overtake him. When it does, he stands for emphasis and engages the bulk of his body in his gestures. I've never seen such vigor in an old person. When Henry speaks, he accents the last word of a comment with the rising inflection characteristic of the French. He involves himself so intensely in his story that he sometimes raises his voice to a scratchy near-holler. All in all, he likes to be interviewed. "Put that in your book," he tells me repeatedly.

Henry says that he began helping out his big family when he was four. He tells me about those childhood duties the first day I talk to him. His job then was to keep firewood in his parents' boarding house, which had twenty-six rooms on two floors and four-man bunks that slept about a hundred men on the third floor. Lumberjacks and mining company workers stayed at the facility. The building had four big stoves that Henry kept supplied with wood.

Talking about tending those stoves brings a sudden realization to Henry, and he looks like a child caught playing hooky. He rises quickly out of his chair. But his larger-than-life gusto collapses like a broken balloon, and he says sheepishly, "I gotta make a fire. I'm gonna get killed. Oh, good thing I thought of that. My wife warned me about that. 'Make sure you keep the fire going. It's going to get cold tonight.'" He stokes up the wood burner in the kitchen. When he sits back down, he is noticeably relieved and begins talking again about his youth.

When Henry was still a boy, he had to go out into the woods and cut small, straight hard maple poles that his parents put under the mattresses on the bunks between the four men.

"Snortin' poles," I say, remembering Swede Intermill's term.

"I never heard that," Henry says.

But he explains that the small maples helped deter fights, which were as common as bed bugs. "You ought to hear the fights I heard," he says. "Holy cripe. *Saatana Perkele*"—he uses the two worst swear words in Finnish, both meaning *devil*—"That's all you could hear." Then, rubbing his throat, he says, "Well, anyway, too bad I got a sore throat."

Henry left school after the sixth grade and went to live on a farm with his sister. He did farm chores in exchange for room and board. He learned how to raise livestock and butcher, how to grow vegetables, and "where the egg came from." He ate well for the first time in his life. When he was fourteen, he returned home and got a job piling lumber at a sawmill. Then he worked as a stevedore, loading hard maple on boats bound for the flooring and furniture mills of the big cities in the Midwest.

About the same time, he got a job with Onesime Dion, a Lake Linden lumberman, chopping down big hard maple trees. The forty-foot logs were used at a copper smelter where they were slid into the molten copper to burn impurities away. The big calluses that Henry got from chopping the maple trees cracked open along the lines in the palms of his hands. He shows his palm and traces the lines, a gesture which reminds him that some Indians he used to work with envied his long lifeline. The cracked calluses opened up so deep, he says, that "you could see the meat on each side." Other lumberjacks told him to urinate on his hands. "I pissed on 'em every night," he says, "and in the daytime I put balsam pitch on 'em. That and piss was good. And in no time my hand got tough and hard. But the first week I pret'near died from pain."

Onesime Dion had two teams of horses to skid the logs out of the bush where Henry was chopping. In a rare moment of modesty, Henry admits that he and a friend couldn't keep even one team busy, even though, he says, "you had a axe you could shave with." When they chopped down the frozen maple, he says, "she'd ring like a bell"—especially the bird's-eye, which he says is harder than plain hard maple. Bird's-eye burned up in a smelter seems as excessive as using vintage wine for mud in the eye.

Such concerns weren't on Henry's mind; his worry was his hurting hands and how to treat them—and this was not the first time that he had taken the urine cure. When he was an infant, his mother told him, he had sores on his face that would scab over. Nothing his mother or the doctors did cured it, which perhaps was a foreshadowing of the way Henry would get along with doctors for the rest of his life. Henry's great uncle Louis, the man who cut off his own toe, knew many Indians and, on a trip to Canada, asked a medicine man what would cure those sores. Henry remembers being told: "He said, 'Make the baby piss in your hand, then wash his face with it.'" His mother did; in three days all the scabs fell off. Henry's beefy hands touch his face softly, although they are so big that they don't seem capable of gentleness. "I had the nicest complexion all my life," he says. "Not a freckle, not a wart, and pink. Oh, a beautiful complexion. I know 'cause the girls chased me all over hell."

Henry chopped down the hard maple for three months, then returned to town to pile lumber at a sawmill. He was paid $27.50 a week for sixty hours of work. "I never seen my check," he says. "My check was going to the bank to pay the eight-hundred-dollar mortgage on the boarding house." He was single and living at home, and those were truly spartan times. His mother would give him fifty cents on Saturday night. He'd take his girlfriend to the show (tickets were ten cents), take her for ice cream afterwards (banana splits were ten cents), and he'd have a dime left over for the Sunday church collection.

Piling wood at the sawmill gave Henry work until he was nineteen. Then he was laid off, so he left for Milwaukee. The day of departure is impressed on his mind yet. He remembers that his mother stood on the front porch and wiped a tear away with her apron. "I can see her yet," Henry says. "She said, 'When you leave here, it ain't going to be home no more. Then you gotta pay for everything, and then you'll know what it is to eat wild cow.' And she was right. I never forgot those words."

A low-paying job in Milwaukee soured Henry on that city, and three days later he was in Detroit, where he married the girlfriend on whom he used to spend the twenty cents a week. They returned home a year later because she was homesick.

Henry interrupts his story to cough and rasp badly. "That tender spot where that lumberjack grabbed me," he explains, rubbing his throat again. "I never took any pills in my life," he complains. He coughs again in great spasms and says it's the Lake Superior catarrh. "There's a lot of people that got that," he says.

"Did it ever slow you down?" I ask.

"I think it was a good thing," he says. "You'd spit all that goddamn rotten stuff out of your system early in the morning. Most everybody's got that, but they don't know what the hell it is, and the doctors can't cure it. They call that a cold. But that's on account of the goddamn Great Lakes here. Damp and sharp and mildew, like, you know—you get all wet. Certain times, not all the time. Mostly in the winter. It never bothered me in the summer. Never spit in the summer. Only in the winter. I've had that, I'd say, about forty-five years, and all my brothers had that, and some of my sisters. My sons never had it. My wife ain't got that at all. She don't spit or nothing."

Again he blames the pills for his sore throat and blames the doctors for the pills. Doctors, he says, "Sometimes they give you the wrong medicine for what you ain't got." At his last checkup, Henry says, he told the doctor that he, that is, Henry, wasn't afraid to die. The doctor told Henry that *he* is afraid to die. "Well, you should be," Henry says he told the doctor. "You guys been robbing the public for years."

Henry talks as hard as he used to work, but his memory is beginning to fail him, and, as with the oxen's hoof, sometimes one recollection will break into a spray of others, like spent buckshot. On one of my visits, I get a good sample of Henry's scatter-gun memory. I ask him about cruising for timber, which he says he did for two lumbermen, Isaac and William Bonifas. Isaac Bonifas was a partner in a sawmill just outside of Lake Linden. His brother William lived in Escanaba. The land Henry helped cruise was between Rice Lake and Mud Lake, southeast of Lake Linden. Henry accompanied Bonifas's cruiser, Chris Erickson. Rice Lake reminds Henry of the Beer Log, a big pine windfall where hikers used to rest on the way to the lake. The Beer Log was where the big timber started. Some of it was elm, which reminds Henry that elm grows the longest root he's ever seen on a tree and tapers down so nicely on the end that you could make a horse whip out of it if it didn't dry out.

"Where was I?" Henry asks.

Cruising timber, I remind him.

He then tells me that he and Erickson entered the bush to cruise by Bissonette's farm, which reminds him how big the Bissonette boys were and how they could lick any opponent in a fight because they knew the scientific way of boxing. The scientific way of boxing is different from a lumberjack's roundhouse, which Henry demonstrates by rearing back with his shoulder and arm as though he's trying to hit the sun. Henry knows about scientific boxing because he took a correspondence course on it from Jack Dempsey's trainer, who taught him that there are only twelve moves with the hands and head—"All the rest is footwork."

"Where was I?" Henry asks.

I have to think. Entering the bush by Bissonette's farm, I tell him.

Starting from that point, Henry searched for the section's corner posts, called *witnesses* in those days. Relating that search reminds Henry how hard it is to keep your bearings in the bush because of "the mind and that short leg of yours." But it wasn't hard to know where you were in this forest because the timber was big, with no undergrowth, and you could see far and line up on a tree. There was a big swamp with lots of cedar, and he and Erickson skirted the swamp and kept it to one side. "A swamp needs it wet," Henry says. On the opposite side of him from the swamp was the big timber—"As far as I could see, oh, it was beautiful."

The trees impressed Chris Erickson—"He pret'near went nuts"—and he was anxious to report back to the Bonifas brothers. The thought of the Bonifas brothers reminds Henry how big the two of them were—both were well over six feet, he says—and how one of them could carry a sixty-gallon pork barrel on his shoulder.

"Where was I?"

Big timber? I guess.

"Oh, it was beautiful," Henry says, and he remembers Erickson exclaiming, "When I tell old Bonifas, he's going to brighten up." Bonifas put a camp in there and cut it. "Nineteen thirty-five it was," Henry says, "or thirty-six."

It is a long, circuitous story. The crux of it is that Henry knew the land well because he had hunted it since he was twelve years old and so was able to guide Erickson. It was the best timber that Henry has ever seen: huge elm; lots of maple, including bird's-eye; and big hemlock and pine. He says the bird's-eye was so good that you could see the bumps in the bark. "And some of those pines was seven feet on the stump," Henry says, "because I had a small ruler under one of my compasses, and I would measure them with that, and it was unbelievable. It was seven feet!"

Henry is beside himself. He gets so excited when talking and is so wrought with emotion that I'm afraid that his oversized heart will burst, especially when he inevitably disparages doctors, a subject that usually provokes him to raise his voice to a near-holler: "I never had a pain IN MY LIFE! Are they CRAZY?"

Sometimes Henry's memory for details from fifty years ago is astounding, but dates are as lost to him as his youth. On one of my visits, he and I spend a long time piecing together his work history.

"What did you do in the twenties?" I ask.

"Oh, they was wild," he says. "Wild. Holy Jesus Christ! Girls found out."

"What'd they find out?" I ask, a bit densely.

"I can tell you're no lumberjack," Henry scolds. "Boy, you were born a long time after me. You heard about the wild twenties?"

He loses his train of thought and, as he does with his sore throat, blames his forgetfulness on the medicine he's taking. "See," he says, "too many pills. I think those pills, they kill me."

When Henry returned to Lake Linden from the city about 1925 or 1926, he walked to Gay, which is fourteen miles east of Lake Linden on Lake Superior. Like so many other Keweenaw towns, Gay has its lonesome smokestack and its stamp sandbanks. Henry worked there at a mining mill for a year. But he became real edgy out in Gay because he didn't like the sound of the waves constantly surging against the shore. To imitate the waves, he makes a sound like a moaning ghost; if he really heard such a sound, it is no wonder he was spooked.

During the early years of the Depression, when he was out of work, Henry lived permanently on a farm and fed his family on vegetables, domestic livestock, and wild game. He says that an old French aunt taught him how to snare deer around the neck; she told Henry that he had to rub the snare with deer droppings to camouflage his scent. He recalls how he used to preserve meat by putting it in barrels and putting salt between each layer. To attract deer to his yard, he used to pour the salty brine from the pork or beef barrels onto a big pine stump in his yard. The deer flocked in. They were "crazy for salt" and licked that stump until it was so smooth, Henry says, that "you'd think it went through a planing mill."

In the mid-1930s, Henry went to work for Bonifas. In the fall of 1936, he went to see a supervisor for Calumet & Hecla Mining Co.—"a big shot who had lots of money and wore expensive suits, pret'near silk"—and landed a job at a coal dock. There he painted a crane that unloaded coal boats. The crane ran along a 550-foot elevated track, called a bridge. Henry took to what he calls "high-climbing" like a bird to the air, and he also painted smokestacks and flag-poles. "I had nerves of iron," he says.

In 1935 he built a sawmill, and in 1938 he bought a section of hard maple near Rice Lake. The maple was characteristic of the Keweenaw, "with hearts about that big," says Henry while making a circle the size of a quarter with his big finger and thumb. He sold the veneer logs—usually the butt log, sometimes the first two logs—to a ladies' shoe heel factory for $120 a thousand board feet. The rest of the logs were milled into railroad ties and sold to the Duluth, South Shore & Atlantic Railroad and the Great Northern Railroad. "You couldn't make enough of them," he says of the maple ties.

As the war approached, Henry operated the crane that unloaded coal boats, working nights at that job and during the day logging and running sawmills. During the war, he had to work two shifts unloading coal boats, so he ceased his logging and milling. When the demands of his crane job eased, he went into Sievi country, northeast of Lake Linden. *Sievi* is Finnish for *nice* or *handsome*, and the area was aptly named, for Henry cut another section of nice hard maple there and sold it for shoe heel stock and railroad ties.

I ask Henry in general about Keweenaw hard maple. "It's the best in the world," he says. Other maple from out of the area that he's seen has yellow stains that are darker than the white wood. "When you run that through a dry kiln, then run it through a planer, you still see that yellow tint in there," he says, "so it don't look good. And birch is the same thing. You gotta be careful. Canada's birch is no good. Nobody buys birch for furniture over in Canada because it's all stained that yellow stain. But here, when you cut a curly yellow birch, boy, that's a beautiful tree. I cut some of them."

Henry's post-war logging in Sievi country succeeded so well that he quit as a crane operator at the mining company in 1948 and put all his energy into logging, running his sawmill, planting trees, and selling Christmas trees, for which he became known as "The Copper Country Christmas Tree King." He sold hundreds of Christmas trees in Detroit every year and got some insights into sales, which, by his reckoning, is a tough job. "Charisma. That's what you gotta have," he says. "You gotta be born with that." The trait, Henry says, is "a gift from God" and gives one "everything it takes" to be rich.

"A man born with charisma can do anything," he says. "Anything at all."

"Do you have charisma?" I ask.

"Oh, sure. Charisma is the gift to sell."

He says he sold seven thousand dollars' worth of Christmas trees every year in Detroit, while "the whole clique" of his helpers could only sell a tenth of that. "I used to tell them, 'You ain't got no charisma. You ain't got no personality.'" He continues what is obviously a pet subject: "The guys I put selling trees couldn't parboil shit for a tramp. That's because they didn't know nothing. They couldn't sell.

They say, 'Twenty-five bucks.' 'Are you crazy?' the guy will ask my seller. 'Are you crazy?' Nobody never asked me if I was crazy, because I have a way about myself. Very sympathetic eyes got lots to do with it. You make your eyes go sympathetic just like a doe that been shot in the side. Did you ever look at her eyes before she died? I did. I sat down on a doe's shoulder once and cried, and I'll never shoot another one. Never did either."

As with hard maple, Henry has nearly a lifelong association with Christmas trees. He started selling them when he was about seven. He remembers vividly the horse, Curly Alice, that pulled the sleigh full of trees. "She was a beautiful beast," he says. "She had long, curly hair. I never seen a horse like that. She must'a come from Ethiopia or somewhere."

Throughout his life, the profit was good from Christmas trees. It costs Henry less than a dollar, plus a decade, to produce a Christmas tree that he can sell for thirty-five dollars. All he has to do is prune the tree twice while it's growing, and once more after he cuts it and before he sells it. Planting twelve hundred trees on an acre of land will yield nine hundred nice enough to sell.

"You think there ain't very much money to trees, eh?" Henry says. "You gotta be shrewd, you wanna make money the way I made it. I learned to lie, all right. Not because I wanted to, but I was only thirteen days in Detroit, and I had to use every way of thinking. I take a tree that you couldn't give to anybody, start pruning it, and, when you look at that tree, you'd never recognize that thing. The price jumped." Henry remembers once having several spruces which dried out and yellowed. "Gold-colored needles," he recalls telling customers. "That's what they call them—gold-colored needles. They sell for a little more." But Henry was too smart to believe he could fool everybody. "You got to watch your bullshit," he says. "You might be talking to a forester." And he wasn't always plain mercenary. "I always lowered the price down when I seen lots of kids, because I was poor one time and I never forgot." He adds on another occasion: "You want to get ahead in this world, your friends are your greatest assets."

The day we talk about Christmas trees, Henry has just been to see his doctor, who told Henry he should retire from the Christmas tree business. Henry responded this way: "I said, 'Are you crazy? My people live to a hundred, and they all work.' I said, 'My health came from the Indians, not from the whites.' I said, 'Hard work kills nobody. If the body is healthy, it will take care of anything, and it will heal itself without you birds around.'"

Henry says that he thinks that the key to his health is "pret'near like an outfit like this"—and he points to a lump as big as a slice of cantaloupe on his arm. "That's why I never took it out," he says of the lump, "because I figure this might be it. It's called the immune system. I got that from the Indian side of my family."

Despite his feeling good and his carping about doctors, Henry allows that he just might retire from the Christmas tree business. Another business that he's ceased to work at is his tree farms. He's planted them all over the Keweenaw and figures that he's responsible for growing thousands and thousands of trees on the peninsula. A good bunch of them surround his house, which is tucked away in a thick copse of evergreens that he planted years ago. Some of them are Colorado blue spruce, several of which are silver. "You come here in the last two weeks of June and the first two weeks of July, you will see the most beautiful trees you ever laid your eyes on right here," he says of the light-colored ones. "All them trees will have silver color, and with the sun shining on the new needles that just come out in that month, it's just like silver, pure silver, coming outta the forge."

Those blue spruce nearly hide his small bungalow, part of which is an old log cabin. The house is a couple hundred feet off the road. From roadside to house, Henry has a snow walk—a wooden walkway three feet off the ground. Entering the property, a visitor climbs up steps by the road, walks along raised planks, and then steps down to the porch entrance. Snow walks are a common way in the north country to cut down on moving snow. Instead of shoveling snow from the ground up onto snowbanks, which get ever and ever higher, Henry now runs his snow scoop along the elevated walkway, in effect pushing the snow off like a plow. Shoveling a path used to take him up to two hours. Now clearing the snow walk generally takes less than five minutes. Not one ever to shy from work, Henry nevertheless is proud of the efficiency of the arrangement. His home, too, has a practical touch—nothing frivolous, all old appointments. The kitchen walls are covered with oilcloth that has been painted white. The stove in the kitchen has a wood burner on one side. "Best range ever made," Henry says. He paid $160 for it thirty-five years ago. Henry brags that it'll take a piece of firewood twenty-six inches long, which is bigger than what his newer wood furnace in the living room will accommodate. Henry's hot water heater stands in a corner of the kitchen near an old sink. His light switches are the old, knobby, round ones.

His small, modest home and its dated furnishings belie what Henry touts as his financial success. He repeats that he retired at age fifty-eight and adds, "If you knew the money I made, you wouldn't believe it. But money don't mean nothing to me. I never spent a nickel in my life. Not because I'm cheap. I gave it away. I got a big family—I helped 'em all out. I never loved money. I never spent any. Money was never good to me because I never enjoyed it. My brother used to say to me, 'You got money! Christ, you don't spend nothing. You still got the first nickel you made.' I said, 'I enjoy seeing the pile grow.' I don't enjoy spending money. Some people enjoy spending. They're broke all the time—look at 'em."

Despite his proclamation of success, Henry's main business, sawmills, started modestly. "Oh, don't think we didn't start poor," he says. "I know what it is—hardship." His first sawmill was powered by a Nash car. The Nash provided 12 horsepower and ran a 12-inch saw blade. The next mill was powered by a Jeep, provided 15 horsepower, and ran a 24-inch saw blade. Then he bought a diesel engine that provided 65 horsepower and ran a 52-inch saw blade. He still has that diesel and mill in the yard by his home, and his sons still use it some weekends.

A sawmill is a precise operation, Henry says. "You see," he says, "when you're sawing with a sawmill, there could be a thousand things to knock the mill going haywire. You gotta be really gifted to fix it. If you're not gifted, that's hard to find that trouble." He talks about correcting what goes out of line by using everything from a plumb bob to a tap with a sledge hammer.

He rises from his chair in the kitchen and beckons me to follow. From his desk in the next room he pulls out a picture album. "You'll see me with my eyes madder than hell," he says. He leafs through the album and finds an old picture of himself with a big elm log that has two railroad spikes imbedded close to the heart. They ruined a saw blade that he had bought for $350. "I made a lot of money off of it before she broke," he says.

In 1937, he bought a bigger mill from a company in New York State, used it for awhile, and then sold it to his neighbor because it was too small and vibrated a lot. "Know what happened?" he asks. "I said to my wife one day while she was making supper, 'Somebody's going to get killed. I feel that way. I think it's that fella running that mill over there.'"

He ran to the neighbor's but the man running the mill had just left. Henry told the man's wife that her husband was running the mill too fast; that when he returned he shouldn't run at maximum rpm's, as he was doing.

"She told me, when I went to the wake a couple days after, that he got killed about eight o'clock that night, after I left. Well, what happened there, the saw was going so fast, she told him, 'Henry told ya, don't run your saw so fast. If it hits anything, it's going to kill ya.' 'The hell with Hank.' No sooner said that than the goddamn saw picked up a knot, and it came and hit him right here"—Henry thumps his breastbone —"It came so fast, she was right there in back of him, and he went down like a log. He died at four o'clock that morning."

Henry sometimes has premonitions like that. He says he has known several times that accidents would happen and people would die. He couldn't, however, foresee one of his own misfortunes. When he was milling logs one time, a piece of wood about the size of a small baseball bat was sent flying by the saw, hit Henry a glancing blow in the side of the stomach, flew outside the open-ended mill, and hit the ground more than a hundred feet away. Something in Henry's insides was damaged, but Henry didn't get it treated for a couple of years, so that on many nights his wife had to push his guts, which bulged out beneath his skin, back into his abdomen. Henry's son, Henry, Jr., who is nicknamed Pucko, was working with his father when the accident occurred and measured how far that piece of wood was airborne after it hit Henry. Pucko says that, just as Henry warned his neighbor about working the mill too hard and fast, he, Pucko, warned his dad about the same thing, but Henry ignored him.

That incident, though, doesn't take away from what Henry sees as other extraordinary experiences in his life. He says he once prayed to God to stop a rainstorm until he had all his dry hay in the barn, and the storm split in two and skirted the edges of his field. He says he knew an old French priest with healing powers, and a French lumberjack, prone to drink, who was a bloodstopper. Henry says he saw the bloodstopper say something in Latin, put his finger on a bad gash in an axman's leg, and stem the flow of blood. Henry guesses that he has premonitions—or has witnessed other people's mystical powers—because he has a lot of faith in God. "I had God with me every day of my life," he says. "From my first breath. Because my parents were very religious and they believed everything, and me, I listened to that and I believed in everything myself. It was easy to make me believe. My mother said to me once, 'See that cow, Henry?' 'Yah.' 'That black cow is eating green grass, and she gives white milk. Can you make that one out?' I said, 'No.' She said, 'God done that.'"

His faith aside, Henry had laid a more earthly foundation for success in the sawmill business years before when he learned how to grade lumber at Bonifas's sawmill in the 1930s. When he first started working there, he was a greenhorn and couldn't tell the difference between hard maple and pine. An oldtimer graded lumber at the mill. "Him and I were big friends," Henry says. Henry asked the man to teach him how to grade. Between what the old man told him and reading books, Henry learned the skill. The old man couldn't read the books that he let Henry read; instead his wife read them to him. But Henry was able to read them, and he says of the old man and himself, "He had no education, and I went to school up to the sixth grade."

Henry learned that hardwood has grades denoting the poorest wood to the most select. He learned that knots, heartwood, cracks, and other defects make for the difference among the grades. The best lumber comes from the white sapwood close to the outside of the logs and low on the tree, where the sapwood has grown over old branches and is clear of knots. The lumber becomes knotty and deteriorates in grade the closer it is to the heart and the higher it is on the tree. Often several grades of lumber are found in one log, Henry says, so the sawyer must know the best way to cut a log to get the optimum grades—and money. "When you're making money, your eyes are open," Henry says. The

97

sawyer's job at the sawmill: assess the log and get out of it the best lumber, take out the taper on the first cuts on each face, and decide how thick the cut should be on each pass through the blade. "There's a lot to that," Henry says.

The mill must be absolutely plumb and square with the world to make certain the lumber is milled correctly. For instance, when milling boards that ultimately will be three-quarters of an inch thick, there is an inch and a quarter between cuts; a quarter-inch is for sawdust, a quarter-inch is for planing, and the remaining three-quarters of an inch is the thickness of standard lumber. The key, Henry says, is to make sure all the machinery is lined up perfectly so that the boards have uniform thickness for eight to sixteen feet, the customary lengths for logs and for lumber. Henry explains that the carriage, which holds the logs and passes them by the stationary blade, must be perfectly parallel to the saw to get a good cut. Otherwise, Henry says, the lumber comes out "thick and thin, thick and thin"—he says "tick and tin"—and isn't marketable.

Henry and I talk once a week. We begin in the winter, when the snow-covered farm fields of the Trap Rock Valley are blown so smooth that they are absolutely without shadow. During the day, the sun glaring off of the white expanse makes it easy to imagine snowblindness. At night, a glowing moon with a delicate aura casts a soft light on the snow and dispels some of the darkness. After I talk to Henry for six weeks, it is early spring, and the slush and mud are like a mire on Henry's long, dirt driveway. Two more weeks, and the Keweenaw is into its characteristically dirty late spring—when snowbanks melt down through the black road sand, which "rises" to the surface of the banks, and, like scum on a pond, looks the picture of unsightliness.

During my last visit, just as we finish talking, Henry promises to show me a sawmill when the weather warms up and the ground dries out. As I ready to leave, Henry jumps up. "I gotta make a fire," he says. "Christ, I'm going to get killed." He stokes up the kitchen stove again; just then his wife comes in from shopping. Big, burly Henry becomes as harried as a crow chased by a songbird, or an eagle chased by a crow. He quavers in front of a woman who weighs less than a hundred pounds and is as slight as a wisp of dandelion seed. "I just made a good fire," Henry tells her meekly. "I was talking and forgot about it. But your fires are going to be hot."

"That's okay," his wife says.

Henry is visibly relieved.

A few weeks later, in early summer, on a cool, drizzly morning, Henry takes me to a sawmill nearby. It is the first time I've seen Henry outside. He steps gingerly; his walk is an old man's walk and not as robust as his talk.

The mill site has three buildings: a dry kiln, a planing mill, and a sawmill. Outside, a man in a forklift loaded with logs bounces along on the wet, bumpy ground to an opening in the side of the sawmill, where he unloads the logs for the sawyer. The high-pitched whine of the mill saw at work carries outside.

Inside the building, which is about seventy feet long and thirty feet wide, three men are working: the sawyer, the edger, and the piler. The machinery runs the length of the building. The carriage that holds and moves the logs has two three-foot-high vertical steel rails, or stanchions, several feet apart. A hydraulic cylinder pushes the log to be cut tight up against these rails. Then two pointed steel pins that are about the size of railroad spikes—"Those are the 'dogs,'" Henry yells at me over the noise of the saw—impale the top of the log a few inches deep and close to the edge and hold it in place on the carriage. The carriage is powered by compressed air and slides back and forth on rails past the fifty-two-inch stationary saw. After several cuts, the sawyer stops the carriage, and a three-foot-long, chain-like apparatus with big barbs goes to work. After the dogs release the log, the chain-like apparatus turns like a chain saw, but slowly. As it does, the barbs catch the log and turn it 180 degrees, so that the cut face is now against the steel rails on the carriage. Once more the hydraulic cylinder pushes the log against the vertical stanchions, and the dogs pierce the log to hold it tightly to the carriage. The sawyer makes three or four more cuts on the new face, after which he turns the log ninety degrees and cuts another new face. Still more cuts, and then the sawyer uses the barbed chain to turn the log 180 degrees to the last uncut face.

The first cut on each face is called the slab; one side is flat, and the other side is curved and has bark on it. Four slabs effectively take the taper out of the log and square it up. That first slab cut on each face is remarkable; seeing the dark bark of the log instantly transformed to white sapwood that looks clean and beautiful is like tearing off the wrapper of a present and seeing what's inside.

On the morning Henry and I watch, the sawyer takes a slab, then three or four boards from each side, and ends up with a three-by-six-inch timber. Each face cut yields one or two clear boards. As the cut moves inward toward the heart, the boards get knotty. Further, each successive cut yields bigger boards; they start out four to six inches wide and end up a foot wide. "The bigger the log, the better the lumber," Henry hollers. "The smaller the log, then you hit the knots."

I look at Henry as he watches the operation. His hands, rarely idle, are sunk deep in his pockets, his eyes look wistful, and he appears oddly fragile and vulnerable.

What Henry calls a "live belt"—a conveyor—takes the cut boards to the man running the edger, which trims the bark off the edges of the boards and makes dimensioned lumber. A man standing on the far side of the edger piles the boards and timbers. The sawyer sits opposite the carriage, close to the saw, in a little glassed-in cabin. He has many levers and buttons to push as he controls the carriage, dogs, and barbed chain. The dog is run by hydraulics, the saw by an electric motor. Depending on the length of the log, it takes four to eight seconds to make a cut and only a few more seconds for the carriage to return to the starting position.

Henry and I watch the operation for a half hour. Sawdust floats in the air like somnolent bugs, even though a blower sends most of it outside, and small wood chips from the logs being cut steadily pepper us. It is a noisy affair, everybody works fast, and the pleasing smell of freshly cut wood fills the air.

Henry and I walk back through the rain to the car.

"That was the first time I was in a sawmill," I offer. "It's fascinating."

"Yah, it's interesting," Henry answers. "Lotta money."

Then we get in the car and drive to his home in relative silence. It is the first time since I met Henry that he has little to say.

John and Barbara Clark

THE MAPLERS

Like fruit trees, milk cows, and artesian springs, hard maple need not die at the hand of humankind to be useful. It gives some of its bounty naturally and with just a little coaxing— namely, its sweet sap, which many people, called maplers, use to make syrup and sugar.

According to Fred Aho, maplers defile the maple forest because a black stain that appears in the sapwood around each taphole ruins the tree for veneer. But maplers aren't interested in wood products, so they carry on a tradition that reaches back to Indian days and remains as much a part of the northwoods as windfalls and widow makers. John and Barbara Clark, who have property and a cabin in the Misery River country in northern Houghton County, each year tap about a hundred hard maple trees, which yield a modest amount of maple syrup. In doing so, they take a little gift from Misery country that, as its name suggests, historically has been hard on the people who try to use its natural bounty.

Swede Intermill, who worked in Misery country for twenty-five years, and Jenny Maria "Jingo" Vachon, whose father in 1900 was one of the first permanent settlers in the same area, tell roughly the same story about how Misery Bay and Misery River got their name. In the 1800s, they say, a small town and trading post at the mouth of the river burned down one winter, and settlers spent terrible months living off the inhospitable land before help came by boat in the spring.

The notoriety of Misery country persisted into the twentieth century. Swede remembers a story that he was told years ago by the wife of one of his loggers. Though he can't recall the woman's name, the story remains vivid in his mind and sobers him yet. The woman's father worked in the area's first lumber camps early in the century. When his wife died, the man took her body into town, buried her, immediately married an eighteen-year-old girl, returned to his cabin, left his new bride with his several children, and went back to work in the lumber camps. The young stepmother sometimes went outside the cabin and, from frustration and loneliness, screamed into the darkness.

More recently, John and Barbara Clark have come to know the hardships of Misery country. Their small, comfortable mapling operation belies their long, arduous struggle, which began in 1972 when, in middle age, they bought 185 acres near the Misery River Road, about a mile and a half south of Lake Superior. They wanted a piece of wilderness, and, over the years, the Misery country extracted from them much sacrifice and toil—rigors that kept them in good physical shape but also have them wondering whether what they've done and learned means anything at all to anybody but themselves.

John Clark, a retired electrical engineer and college professor and administrator, is trim, white-haired, and was born in 1918. Barbara Clark, a retired high school biology teacher, is tall, dark-haired, and was born in 1926. When I first met John and Barbara, they invited me to their cabin to watch their spring mapling operation. It took four years for me to take advantage of their offer, and on a cold, sunny day in late March I drive southwest of Houghton to the crossroad town of Toivola. *Toivola* is Finnish for "place of hope," and it is but seven miles from Toivola west to Misery Bay. About a mile and a half short of there is the trailhead to John and Barbara's cabin, which is landlocked a half mile from the nearest plowed road. In the winter, the only way to get to their cabin is on foot. I put on cross-country skis and a day pack, loop extra boots around my neck, and set out to their cabin.

My arrival is timely, comic, and imprudent. I come to a hilltop from which I can see their cabin, just fifty feet beyond the South Fork of the Elm River. The hill looks steep, so I sidestep partway down. Then I align my skis with the slope, and push off. It is a mistake. The snow is hard and icy, and I zip unsteadily toward a small footbridge at the bottom of the hill. Somehow I make it across without plunging into the stream. Eight feet from the cabin, I have to fall to stop. My boots hit me in the face, my chuke falls off, one chopper fills with snow, and my skis tangle. I look up at the cabin. John and Barbara are gazing out of a bay window, half-surprised, half-dismayed.

They come out, and John says, "Your timing is perfect. We were just talking about you."

Barbara adds, "Yeah, I was saying how I hoped you had enough sense not to ski down that hill."

John and Barbara invite me into the cabin. We sit in the dining room, where an old Silver Oak wood stove casts welcome warmth, for the temperature outside is only in the twenties—too cold, John says, for the maple sap to run. But he has about sixty gallons from earlier in the week, just enough to show me how to make maple syrup. I don't know much about the process but, in preparation for my visit, have read an old but informative book, *The Maple Sugar Book*, by Helen and Scott Nearing.

The Nearings say that the sap flows mostly in the spring before the buds come out. A tree with leaves is "dry," but, when it is leafless and dormant, the sap flows. The Nearings quote naturalist John Burroughs, who says the early-spring sap, "like first love, is always the best, always the fullest, always the sweetest." Later in the spring, just before the buds come out, the sap is less desirable, and syrup from it is sold as a commercial sweetener for, say, baked beans, instead of table syrup. Still later in the spring, when buds begin to come out, the sap darkens, becomes even less sweet, and the syrup from it tastes like burned bacon. The budding of the trees, then, ends mapling.

The Nearings say that the sap flows best when the daytime temperature reaches between thirty-three and fifty degrees Fahrenheit and the nighttime temperature dips below freezing. Once again, they quote Burroughs: "The moment the contest between the sun and the frost fairly begins, sugar weather begins, and the more even the contest, the more the sweet."

When their book was published in the early 1950s, the Nearings lived in New England and made a living by making maple syrup. They took their mapling seriously, warned against lukewarm interest, and quoted with devotion this unattributed advice: "Do not suffer a hand employed in your sugar camp to ever carry such deadly weapons as guns and rum bottles, nor articles so destructive to success as cards, dice, dominoes, and novels. You must watch and work, and then you need not doubt success. Sugar-making is pleasant, healthy, hard work. A camp is no place for lounging."

John and Barbara explain over coffee that their operation is a hobby, not a business. They get just enough syrup to keep three generations of Clarks supplied with maple-sweetened pancakes and waffles. The year before, John made fifteen gallons of syrup.

Shortly, John and I go outside, fill several five-gallon jugs with water, and lug them to a building called a sugar shack, which is a couple of hundred feet from the cabin. The sugar shack has a cement floor that has absorbed the cold all night. I have on two pairs of socks, one cotton and one wool, plus insulated boots, but soon my feet are cold. John is wearing an orange sweatshirt with a hood, khaki pants, boots, and a red cap. When he takes his cap off, he shows thin, white hair that is stringy, like his eyebrows, which are partly hidden behind dark-rimmed glasses. His dark blue eyes are striking, and he has a soft, even voice that is paced and thoughtful.

Jutting out from the north wall of the sugar shack stands John's evaporator, which is used to boil most of the water out of the sap and render it into syrup. To the left of the evaporator is a shelf with a twenty-gallon, galvanized oblong tub, which sits on a platform that is five feet high. The tub holds sap, which will drain into the evaporator. On the floor are two big plastic tubs containing more sap that John will ladle with a dipper to the elevated feed tub. Opposite the evaporator, on the south wall, is a stack of firewood. On the east and west walls are shelves piled with tools. Everything is orderly, the sign of someone who knows manual labor.

The evaporator, which is about four feet high, two feet wide, and six feet long, consists of a large firebox that is crowned with a six-inch-deep, light metal tray. This tray, which holds the boiling sap, is like a dam on a river: the feeder stream is the trickle of sap that flows by gravity through a hose into one end of the evaporator; the dam's headwall is a float that maintains the level of sap at about two inches; the dam's spillway is a spigot at the other end of the tray that drains off the concentrated sap, or syrup.

Getting the evaporator fired up is a ticklish task, and John's procedure is of his own devising. "I never saw anyone else start one up," he says, "so I just figured out a way to do it." He begins a long explanation of the boiling process.

John can't simply make a fire and turn on the sap flow; before the sap would fill up the tray, the light metal would buckle from the heat and ruin the boiling tray. Similarly, John can't fill up the tray with two inches of sap, make a fire, and begin boiling; all of the sap would become syrup at the same time, and he'd have to try to drain the tray all at once; some of the sap probably would turn to sugar before he had it all drawn off; even if it didn't, once again he would end up with an empty tray that would buckle from the heat.

To solve these two problems, John fills the evaporator to the two-inch level with water, fires up, and begins to let the sap flow in on the feeder end. The sap, John says, slowly pushes the water to the other end of the evaporator, where it drains off through the spigot. When John judges that the evaporator is filled with sap and no water, he will turn off the spigot and boil just the sap. The dilute sap trickling in is controlled by the float and will compensate for the loss of fluid to evaporation and maintain the two-inch level. And, as more sap flows into the tray, or dam, it will push the boiling sap ahead; as the boiling sap courses its way through the evaporator to the spigot, or spillway, it becomes more concentrated. After some hours of boiling, John says, the sap near the spigot, which has been in the evaporation tray the longest, will approach the consistency of syrup. Then John will open the spigot and let the fluid slowly drain into bottles. Now the float will let the feeder stream flow faster to compensate for both the evaporation and the outflow and, like water seeking its own level, will maintain the two-inch depth. At this point, the evaporator, like a self-fulfilling prophecy, maintains itself—there is as much sap going in, John says, as there are steam and syrup coming out—and the slow process continues until quitting time. Then, John will let the feeder tub empty of sap, fill it with water, and let the water push the last of the sap through until only water remains in the evaporator. He then can let the fire die out, use the water to clean the boiling tray, and later finish off the syrup on a gas burner.

The boiling tray that sits on top of the firebox is divided in half across its width by a metal partition. The three-foot section on the feeder end has a corrugated, or fluted, bottom. The flutes are about five inches deep, a half inch wide, and about two inches apart. They expose the sap to more surface, and thus to more heat, and John says that 80 percent of the water is boiled out of the sap in this first section of the evaporator.

The sap in the fluted section flows through an opening in the partition into the front half of the evaporator, the bottom of which is flat, instead of fluted. This flat-bottomed section of the evaporator is divided into four equally sized compartments about six inches wide, six inches deep, and three feet long. The compartments are connected by small holes that allow the sap to flow from one compartment to the other. These openings are on opposite ends of the compartments, so that this section of the evaporator is like an old river or a switchback mountain trail: the sap makes many "S" turns as it flows along to the spigot at the end of the fourth compartment. This zig-zag journey through the four compartments lengthens and slows the sap's journey through the evaporator and isolates the sap in various stages of dilution. As the sap flows slowly through the four compartments, it becomes more and more concentrated. The sap in the first compartment is still mostly sap, clear, and only slightly sweet. The fourth compartment is where the syrup is nearly finished and drawn off through the spigot.

After John explains the boiling process, he puts a hydrometer in the tub of sap that will feed the evaporator. It registers 2.75 percent sugar. The sugar content of maple syrup is rarely above 3 percent, he says, although up to 12 percent has been recorded. Making syrup out of maple sap is like making a mole hill out of a mountain or a puddle from a lake; John's sixty gallons of sap will yield only about two gallons of syrup. The sixty gallons that he has from earlier in the week is enough to "just get the evaporator up to speed, then start shutting it down," he says. It's been cold or the sap wouldn't keep, he adds. In warm weather it goes bad in a day or two.

After John fills the evaporator with water from the jugs that we've carried to the sugar shack, his first job is to clean the evaporator. The year before, he gave it a good scrubbing with phosphoric acid, but now spider webs and dust have accumulated. He cleans the evaporator with acid, he says, because "there's hardness in sap like in hard water. When cooking, it tends to make hard water deposits like in a tea kettle." There's also a fine "sugar sand" that boils out of the sap. Some people filter it out; he lets it settle out. The water in the evaporator soon steams as John sponges out the compartments and collects debris with a small sieve. While doing so, he talks about the sap flow in maple trees.

The sap carries the energy to help the tree grow, he says. "Water is important at the beginning because the leaves and buds are dehydrated." Moisture is stored in the tree and flows on a warm day; it stops flowing on a cold day. The alternating cold nights and warm days of early spring are ideal for mapling. Sap flows at other times of the year, John says; some people tap in the fall, but that is rare. Even in the summer, there's sugar in the tree, he adds. "Pick a twig in the summer and taste it. If you concentrate hard enough, you can detect a little sweetness."

I arrived at the Clarks' cabin midmorning. By the time we have coffee, haul water, and clean the evaporator, it is noon, so we go to the cabin for a lunch of cheese, carrots, celery, peanut butter, pears, and cookies. John explains that a typical mapling day isn't so leisurely. Normally he is at it before breakfast and has lunch on the job.

At 12:40, we are back in the sugar shack, ready to start. The fire has died, so John again fires up the evaporator. He squirts fuel oil on the kindling. "My insurance," he says. "Barbara thinks that's cheating." The door to the firebox carries the words "King Evaporator" and a big dollar sign. Above the evaporator, which he paid seven hundred dollars for secondhand, hangs a big canopy of sheet metal. It is suspended on cords and, when lowered, covers the evaporating pan and carries the steam to a roof vent, keeping the room clear of a fog-like cloud. The canopy has big, round styrofoam pads on the two outside corners in case a person bumps them. John is deliberate and inventive—he rigged up an electrical system for the cabin out of car batteries and made his own temperature gauge to test whether the sap has become syrup—and I wonder if those cushions on the canopy are there because of foresight or hindsight.

I say nothing, however, and watch John stand by the evaporator and wait for the steaming water to boil, a critical juncture in his mapling process. John's method for determining when the sap becomes syrup depends on the boiling point of water. Research has shown that maple syrup boils at 7.1 degrees hotter than water does. Because the boiling point of water depends on elevation and barometric pressure, John brings water to a boil in his evaporator, puts the sensor of his homemade gauge in the water, and sets on zero the meter that measures the temperature in tenths of a degree. With that relative setting, John will know he has syrup when the meter reads 7.1.

Now, though, the water has just begun to boil. John takes its temperature and sets his meter, then begins to let sap drain in and water drain out. It's a judgment call to determine when the evaporator has only sap in it because the water and the sap mix a bit at their interface. But John knows how much water he put in, so he carefully watches the sap level in the feeder tub. When he figures the sap has replaced the water in the evaporator, he will close the spigot and begin to boil the sap. He says that you can see the demarcation between the water and sap better at the tail end of the process—"when putting water in to push the last of the syrup through." The reason: as the sap boils down, it becomes the color of amber, and the difference between the dark sap and the clear water following it is obvious.

While John waits for the water to drain off slowly and the sap to follow, he talks about the hill I was indiscreet enough to ski down. A spring gurgles out of the hillside just off the trail, so even in the summer the trail isn't dry enough to bring a truck down. That wetness has caused him much frustration and work, so that, for instance, John had to have concrete mix dumped at the top of the hill, which he then pulled down to the work site by hand in a dray. All fifteen tons of it. "That didn't happen in one afternoon," he says.

As he talks, John slowly paces around the sugar shack. "The steam doesn't smell like there's much sap in there yet," he says. "It won't hurt if I boil water for awhile. It'll just boil off. It gets a little boring sitting around waiting." He lowers the steam canopy. The level of sap in the feeder tub drops slowly but steadily. The sap drains in at a trickle; the water drains out at a trickle. Soon, John judges that the evaporator is full of sap. "This is it," he says. "Now let's see if we can make this chase its way around." He shuts off the drain, and now the sap flows in to replace, not water, but the fluid lost through evaporation. We must wait until the sap near the spigot nears 7.1 degrees. "It'll be a long wait," John says.

He stokes up the fire; then we go outside. The meadow where his cabin, sugar shack, and other outbuildings are located is dotted with evergreens. The maple stand that he taps is on a hillside about two hundred feet from the sugar shack. Paths for collecting the sap have already been made earlier in the week. John has tapped 110 trees this year. Each taphole is seven-sixteenths of an inch in diameter and two and a half inches deep. A plastic spile is inserted into each taphole. In some places, John has single taps; blue plastic bags hang from those spiles to catch the sap. On one part of the hillside, though, all the taps are joined together with plastic hose and fittings. The hoses catch sap from all the trees in that area and drain into one line, which in turn drains into galvanized garbage cans on the bottom of the hillside. On this day, the temperature is twenty-five degrees, and just a little sap is frozen on the bottom of his garbage cans; the bags are empty. On a warm day in the mid-afternoon, John says, the connected taps make for a tiny stream of sap that flows steadily into the garbage cans, instead of just dripping. He carries the sap in five-gallon jugs to the shack. "I could make life a lot easier on me if I had better means of collecting sap," he says.

John's sugar bush has two drawbacks. First, it faces northeast instead of south, so the sun doesn't warm it as quickly as it might. Secondly, his land is only about a mile from Lake Superior. "The lake is too close and the land doesn't get the rapid temperature changes you get farther inland—in Lower Michigan and northern Wisconsin—where you have mapling going on," he says. Those less pronounced temperature swings retard sap flow, he repeats, because ideal mapling weather is frosty nights and warm days.

In northern Wisconsin, he says, railroad tank cars are loaded with syrup bound to be bottled in Vermont, which is world-famous for its mapling. "It sells better that way," he says. Such commercial operations never have enticed him into a larger operation. "This much is fun," he says of his hundred taps. "It's a pretty iffy business. It varies from tree to tree and season to season." He knows of one man in Misery country who had between ten thousand and fifteen thousand taps and needed fifteen hundred gallons of sap just to start up his evaporator. During one bad year, "He couldn't get that much ahead before it spoiled," John says, and only made three hundred gallons of syrup. The man is no longer in the mapling business.

We leave John's sugar bush and return to the shack to stoke up the fire in the evaporator. John's homemade gauge is barely above the boiling point of water. The gauge disappoints him a little. He says the material he used is like thin aluminum; it is too sensitive and cools and warms like an overly moody person. "You can almost watch the bubbles go by," he says.

John now begins his routine for the afternoon: pacing slowly, ladling sap from the containers on the floor to the feeder tub up high, checking his float, checking his temperature, raising the canopy, skimming scum off the boiling sap with a small sieve, feeding the fire, lowering the canopy, and then pacing slowly and waiting. He always seems to have an objective to his pacing, like showing me a tool or piece of equipment as an illustration to his explanation of mapling, which comes sporadically, when he's not busy, and stretches through the afternoon like a good book.

After some minutes, John again raises the canopy. Steam fills the room. He takes a big ladle and tests the sap. "Boy, there's no real color in there yet, is there?" he says. He sips the little bit in the spoon to taste for sweetness. "If you think real hard," he says, "you can almost imagine it." He lowers the canopy and says, "I'm going to get that fire going a bit more vigorously. The best syrup boils really fast." He explains that when the sap simmers slowly, its natural sugar breaks down into two simpler sugars that aren't as sweet. For that reason, John likes to use relatively small firewood because, just as the fluted section of the evaporator exposes more sap to the heat, small wood has more surface area to burn. He says it's a fast fire but a hot one. He reiterates that his is a simple process. Really modern operations have computerized evaporators with many heat sensors, conveyors, blowers, and controls. Compared to that sophistication, John says, is the ancient Indian practice of putting the sap in hollow logs and adding hot rocks to make it boil and turn to maple sugar.

Again, John raises the canopy to check the sap. The steam feels warm on the face, and I stand close to the evaporator to get its full effect. I can't smell sweetness in the steam, but the Nearings say there is some sugar in the steam because after a long day of boiling, the person tending the evaporator gets sugared eyelashes and sticky clothes and hair.

With the fire in the firebox really cooking now, the floor of the sugar shack is warmer and my feet are no longer cold. John, who hasn't appeared uncomfortable at all, opens an outside door. "We may as well humidify the outdoors," he says.

He checks the temperature gauge. It is 1.0 degree up from the boiling point of water. The sap is boiling wildly. The froth in the flat-bottomed, compartmentalized section almost reaches to the top of the dividers. John walks to his

114

cabin and comes back with butter. He drops a dab in the boiling sap; the foam instantly recedes. In older times, he explains, maplers tied a chunk of suet over the evaporator. When the foam reached it, the fat melted into the sap and automatically kept the foam in check.

John lowers the canopy once more. I am disappointed; I like the steam on my face. He ladles more sap from the floor containers into the feeder tub. While doing so, he says that, when the season is over, he cleans all his equipment, bags, tubes, spiles, tubs, jugs, pails, and evaporator. "Usually by then you're kind of tired of mapling," he says. "It's time to get on to other things."

I ask him if tapping the maple trees hurts them. Yes, he says, but the hole heals quickly. However, a black streak forms in the area around the taphole and never goes away. If you cut down a big old maple tree in a sugar bush and slice off a section at tap height, which is about breast height, you can tell where it was tapped down through the years, he says; the stains show in the growth rings like tattoos. Those, I think, would make the trees useless for veneer or good quality lumber, and I remember Fred Aho talking disparagingly of maplers.

John explains that a sugar bush is different from a forest that is managed for timber. The more the branches and the bigger the canopy, the better the tree is for sap production. Therefore, sugar bushes are thinned more than forests that are managed for saw logs so that they grow more branches and catch more sunlight. John says, "That's one of the few things everyone in mapling seems to agree on—trees that have a large crown are likely to make more sap and sweeter sap." Oddly, another truism maplers sometimes propound is that when a maple tree is slightly stressed, resulting from less-than-ideal growing conditions, the sap is sweeter—"like they claim the sweetest apples come from the tree that's had a hard time," John says. Stress can come from cows grazing, roots under a road, or salt spray, he says. Maple trees along road sides and in parks are sometimes prodigious producers of sap, he adds.

John raises the canopy. Steam envelopes the room. I stand close to the evaporator, happy again to feel the warm steam on my face. My sense of smell is poor, though, and I still can't detect any maple smell in the steam. John spoons out some sap from the four compartments. The first, the farthest from the draw-off valve: "Nothing there to speak of. Boy, it's slow." The second: "There's a little there." The third: "Tastes good." The fourth and last: "Oh my god, yes."

The foam is brown and scummy. John adds a dab of butter to calm it down, then skims it off, along with dirt and debris rising to the top. He checks the gauge; it is up two degrees. I think of the Nearings' compilation of advice on how long to boil the sap. One bit of folk wisdom that they picked up advised: "Boil until it's done."

John stokes up the fire and lowers the canopy. He says there are many ways to make mapling more efficient: reverse osmosis procedures that make concentrated sap like concentrated fruit juice; piping systems that pre-heat the sap before it gets to the evaporator and also capture the distilled water from the sap; apparatus to take care of the sugar sand problem; evaporators that run in a vacuum and run by their own steam. He is content with his modest setup.

2:00, after an hour of boiling sap: 2.6 degrees. Slow progress. "Yeah, we'll just get this thing going good and it'll be time to put on the brakes," John says. "Everything will happen at once and it'll be time to draw off."

John built his sugar shack four years previously. Much of the construction material was brought down the hillside by hand. Once John drove a truck in via a circuitous back way— then on the way out had to winch the truck back up a shallower hill than the one I skied down. He progressed a few inches at a time. It took all afternoon.

John opens the evaporator and chucks in wood. The fire crackles. "Pine," he explains, without my either noticing or asking.

2:20: 3.0 degrees.

John says that his 225 acres was logged before he and Barbara bought it, and now most of the timber is small. The maples that he's tapping, for instance, are all about a foot in diameter. He tries to tap only hard, or sugar, maple but gets some red maple, too, because he has difficulty identifying red maple without the leaves on it. You can also tap box elder, which, he says, the name notwithstanding, is a maple. When John and Barbara first bought their land in 1972, the elm on the property was dying from Dutch elm disease. A forester told them to cut it before it was all ruined and worthless. John and Barbara hired two lumberjacks who were trying to make a go of an operation using horses rather than machines to skid the logs out. "When they pulled out, you could tell they had been here," John says, "but two or three years later you had to look twice. If you weren't looking for it, it wouldn't come to your attention." The elm logs yielded about two thousand dollars for him and Barbara.

A crunching on the snow comes from outside. Barbara appears in the back door with cups and a thermos of coffee, then leaves. 2:27: 3.2 degrees. A cup of coffee later: 3.6 degrees.

"Maybe it's time to look at the clarity again," John says. He raises the canopy and spoons some from the first compartment. "About as clear as can be," he says. Second compartment: "Well, that's getting a little color." Third: "A little more." Fourth: he offers me a taste; it is dark and sweet. John lowers the canopy. Steam condenses on it, and water drips to the floor.

More crunching of snow outside. "Barbara," John says. His understatements are always to the point. I go outside. Barbara is pulling a little sled with a chain saw on it. I join her, and she explains that there are windfalls that must be

117

removed from a cross-country ski trail near where John has tapped trees. We reach the site, and Barbara picks up the saw, chokes it, and starts it fluidly and expertly. Soon sawdust is dirtying the snow as Barbara lops off firewood-size pieces from a six-inch tree. Another twelve-inch windfall is nearby. She saws four-foot-long pieces off of it, and, as she does, the saw binds for the first time. She jerks the saw loose, then cuts from the top until she can see the blade is going to bind, deftly pulls the saw away in the nick of time, and then cuts from the bottom of the log up to the top cut. Each piece falls away nicely. She is as smooth and precise as a dancer. When she has everything cut, she stops the saw, puts it on the snow, and we load the small pieces onto the sled. I ask her why she cut the twelve-inch tree into four-foot lengths and the six-inch tree into short pieces. She explains that the smaller windfall is maple and will make good firewood; the larger one is elm and dead. Were she needful of firewood, she'd take the elm, but with many years' supply of firewood, she tosses the elm off to the side. She has cut it in pieces to help it decompose. She says of the windfalls: "Wind and ice raise havoc with the trees that aren't too strong."

I ask Barbara how she got so handy with a chain saw. She answers that she and John belong to the Michigan Forest Association, which has a program for landowners that includes instruction on using a chain saw. "That and lots of advice," she says. The advice came from her neighbor, John Rauvala, a retired sawyer who worked for Swede Intermill. He taught Barbara how to work in the woods, operate the saw, and maintain it. "I was apprenticed to a highly skilled and experienced logger," she says, adding that working with a chain saw "can be extremely dangerous. It's a wicked weapon when it's running. It can cut through flesh and bone easier than it cuts through maple. I had one squeak. I cut my knee. I had to have it sewn, but it wasn't serious. It wouldn't have happened at all if I would have had safety chaps on."

While she pulls the sled with firewood back to the wood-pile, I put the saw away in the tool shed, then head back to the sugar shack. White steam from the sap and dark smoke from the wood fire pour out of their respective stacks. 2:55: 5.0 degrees. The boiling sounds like a distant waterfall.

3:07: 5.3 degrees. The sap is really cooking now. John has ladled all the sap in the standing containers into the feeder tub, which is already half empty. He puts a box under the draw-off spigot and puts a Jim Beam whiskey bottle on the box and a funnel in the bottle. "We have to watch very carefully this doesn't go dry," he says of the feeder tub. He explains once more that he will push the sap around the evaporator and out of the spigot with water. And he explains again that this is a leisurely day. When he's up before breakfast and works until dark, there is a steady trickle of syrup for two or three hours. "It's really intriguing," he says.

John has tried all kinds of methods for determining when to draw off the syrup. One gauge measures the density of a solid in solution, like sugar or salt in water. Another measures how much a solution refracts light. He shows me old steam pressure gauges that he's made and tried. His training as an electrical engineer has inclined him in favor of his temperature gauge. "I feel more at home with electrical circuits than I do with steam," he says.

3:24: 5.7 degrees. John explains that he will draw off the syrup before it reaches that critical 7.1 degrees. He'll draw off at 6.9 or so and finish the syrup on a gas plate with a heavy aluminum kettle to better control the temperature. The evaporator's light tin easily scorches the syrup.

3:29: 5.8 degrees. John raises the canopy, but not all the way up. Walking around it, he bumps his head and, in doing so, solves a mystery: "As you can see, I bumped my head on the corner of this thing when it was partway up and decided to do something about it," he says of the styrofoam pads.

3:35: 6.1 degrees. John raises the canopy and spoons out the liquid. Each container is decidedly colored. He lets me sample the sweetness. After I test the last two compartments, I feel the stem of the spoon and it is sticky.

John draws a pan of syrup off from the last compartment. He wants the syrup by the spigot to circulate faster so his gauge can measure a bigger sample of the compartment's contents. Then he pours the sampling back in.

3:44: 6.6. The color now looks dark. John waits a minute, then slowly draws off enough to fill the first 1.74-liter Jim Beam bottle. The temperature gauge drops down to 5.5. It's about three hours since we started pushing water out of the evaporator with the sap.

By 4:00, the feeder tub is empty, and John pours in water to complete the water-sap-water cycle. Barbara comes in.

"How's it going?" she asks.

"We're just starting the water chase," John answers.

"How much did you take off?"

"One bottle."

"Oh, it looks good, though."

John says he needs more water, and Barbara and I go for it. The oak barrel that catches the spring water constantly overflows out of a hose; there also is an incoming hose and a spigot and hose to draw off water. All of them are encased in ice. Barbara grabs one, tries to yank it loose from the ice, and breaks it in half; it is the feeder line. "Oh, oh, that's terrible," she says before going into the cabin to get some duct tape. We repair the hose as best we can. My hands ache from the cold water and colder air. Our work is makeshift at best, but effective: the splice leaks only a little; the pressure is still sufficient to draw water out of the barrel.

We haul water back to the sugar shack. John has had trouble with the float and the water feed and is drawing off some of the syrup early. "I'll run the risk of taking it off a little early rather than burn it," he says. "I changed the float probably too far, the level went down, and I had trouble getting it back up. And then the thermometer wasn't really buried, so I couldn't tell my temperature."

He has drawn off a pint plastic jug of syrup for me. "It's close to being finished," he says. I decide to try to finish it off at home on the stove and use it on buckwheat pancakes. Shortly, John has three and a half Jim Beam bottles filled. Two contain dark syrup; the liquid in the other two is light like tea. Barbara enters with a fifth of Jim Beam whiskey.

"Did you tell him?" she asks me.

"No," I say.

"Tell me what?" John asks.

"Maybe you should have some of this first," Barbara says, proffering the whiskey, and then explains about the water hose.

"I'll be darned," John says without a trace of ill temper. "It must be rotten."

Barbara then tells me to put snow in my empty coffee cup. I go outside, follow her instructions, and return. She pours in some hot maple syrup and a good shot of whiskey. "We call them Beamcicles," she says. I sip and realize that now writer Robert Traver has nothing over me. In "Testament of a Fly Fisherman," he said nothing tastes as good as water from a cold trout stream and whiskey. I disagree. A drink of whiskey and snow and maple syrup is an unmatched delight. I have two and change my mind about using the syrup for buckwheat pancakes.

After our drinks, John lets the fire die down and begins to clean up the evaporator. Water is still in it, and he skims off the scum and debris, which includes evergreen needles and dead flies. "Some people say without the bugs you don't have the proper flavor," he says. "They don't drink much, in my opinion." While he works, Barbara and I talk. She says that the weather forecast is for temperatures near zero for the night, and that it won't warm up much during the day, which means the sap won't flow tomorrow either. John moves around us. He bumps his head again on the canopy. "See why I put those things there," he says. He has the water out of the evaporator now and sponges up what won't flow out.

Then he siphons the water out of the flutes in the rear of the evaporator. Barbara interrupts our conversation when something on the floor distracts her. She bends down and picks up a small snake skin lying on the floor. "You can never tell what you'll find out here," she says. "We live close to nature." She leaves to get dinner ready. John shovels the ashes out of the firebox onto the snow outside. He shows me the spot on the evaporator where he burned syrup the year before. It is black and buckled. We finish about six o'clock. We've been at it since mid-morning. "Maybe you can begin to understand why the price of maple syrup is rather high," John says.

We carry the four bottles to the cabin, where Barbara is preparing dinner on a gas stove. John and Barbara also have a gas refrigerator and a gas heater to supplement their wood heat. Three propane tanks that weigh three hundred pounds get them through the winter. The tanks are delivered to the top of the hill. John takes them down to the cabin on a sled. Barbara says, "We used to have a gas man who would deliver them down the hill and leave the bill on the door, but times have changed." She goes to the refrigerator and makes an immediate comment on herself: "Talk about your city gal opening up your gas refrigerator and wondering why the light doesn't come on."

She serves up a meal of baked potato, meat loaf, and salad. For dessert, we have strawberries with homemade biscuits and homemade plain yogurt. Outside it is still light, and the chickadees are in the bird feeder. A tree close to the feeder has silver and gold bells hanging in the branches. Barbara says they are from Christmas. She and John like to hear them tinkle in the wind, so they leave them up all year.

After dinner we sip coffee and talk. We talk of mice that eat hand soap. John says, "I always envision two mice bubbling away trying to talk to one another." We talk of porcupines that eat everything. John and Barbara can hear them at night chewing on the barn. She talks of "demolishing them" with a shovel. She says they're "gruesome and deadly look-ing" upright, but soft and furry on the stomach. We talk of nature. "No matter what you try to do, nature comes one up on you," Barbara says. "It makes you humble." We talk of the electric lights that they found in their cabin when they first bought it. "We thought that was incongruous and hid-eous and couldn't wait to get it ripped out," Barbara says. "It wasn't exactly a backwoods thing. But if pioneers had lights, they'd have used them. It's a matter of what works. But it was a while before we succumbed to a gas refrigerator and a gas stove."

Then we talk about what Barbara calls her "big conver-sion." She speaks of herself and her thinking because she is the dominant political voice in her and John's relationship. The gist of her story is simple: how working on their Misery property changed her—from a strict environmentalist who opposed logging, to a staunch advocate of managing forests through logging. John and Barbara first learned about the Misery site in 1971. For several years they had been looking for property. One day a colleague of Barbara's mentioned that her family had a camp. Barbara remembers the day well: "She said they didn't ever come out here anymore, so they supposed they might be interested in selling it, but 'you wouldn't be interested in that.' I said, 'Why not?' She said, 'Well, you can only go so far and then you have to cross a creek. It's an old Finnish log cabin that's all tumbled down. And it's a hundred and eighty-five acres.' Everything she said turned the lights on brighter.

"So finally I wormed out of her where this place was, and when we came, the porch, which was on the front side here, had fallen down and was a pile of rubble. And there was a back shed and it was leaking. The roof leaked fearsomely onto a wood stove that was there. It was all rusty. And the door frames were screw-jawed, and the foundation was crumbling.

"But there were some small spruce trees and balsam trees out there. They're a lot bigger now, but then they were just small Christmas-tree style things, and one of the kids hollered at us, 'You should come and look at this.' And there was a fawn curled up under one of those trees. And as we walked in—we had to walk in the whole way because the road was in such soggy shape that you couldn't drive in—the violets were just blue, just a solid bank of blue on either side of the road. It just really wiped us out. There was a little, teeny, rickety, two-plank bridge over the creek—you hardly knew whether it was going to last until you got across. And a big log was across the road, down, so you had to climb over the log. It was quite a challenge just to get in here. But it was really so remote, and so far from neighbors, and it just hit our fancy."

They bought the parcel and about five years later bought forty more acres, which were known as the Rinkinen farm. The original plot was mostly forest; the Rinkinen farm had a barn, a stone well, and an old apple orchard; the rest was field and woods.

So two city folks, he from Indiana, she from southern Michigan, had their land in the wilderness. They immediately started restoring the cabin. They had to tear down porches and sheds. The floor was so out of level—Barbara says "a square block would pret'near turn over"—that they had to jack the cabin up, replace the foundation, and fill in the root cellar beneath it. They had to build a bridge with big cedar logs that they cut quite a distance away. Then they chained one end of the logs to an axle with two wheels, left the other end dragging, and pushed the wheels down the creek bottom. When they rebuilt the foundation, they had to carry concrete blocks, one in each hand, down that devil of a hill. All the work was done, Barbara says, "by just the two of us, and the kids alternately, intermittently, and occasionally."

124

John remembers with a grimace the job one weekend of hauling six plywood sheets into the site during the winter. He rigged up a "kind of a sled arrangement" that carried one end of the boards while he lifted the other end and pushed. He had to snowshoe in and could skid in two sheets at a time to the top of the hill. Then he carried them one-by-one down the hill to the cabin. "My leg, I had banged it in town," John remembers. "The flour mill fell on it off of the table while I was putting linoleum down in the kitchen. And my leg was pretty sore from where that thing hit, but it seemed to be getting better. Then the next weekend I brought all this wood in here, and the next Monday it ached so. Well, anyway, I went to see the doctor. 'What in the world is hurting my leg?' He had an x-ray, and it was broken."

"Nobody can say you're lazy," I observe.

"I am now," John says.

"We've aged considerably in the last five years," Barbara says. "I'd say partly because we killed ourselves earlier."

But they had their wilderness, and Barbara was a fierce wilderness proponent. She was instrumental in the 1960s in getting the second wilderness bill passed in Congress. The bill created wilderness areas in Shenandoah National Park, Isle Royale National Park in Lake Superior, and a few places out West. Barbara was partly responsible for the part of the bill involving Isle Royale National Park. She made a map of the island, which is about fifty miles from the shores of the Keweenaw Peninsula, and showed what spots should be campsites and what areas should be left untouched. She brought the map to Washington, D. C., and testified about the plan before a congressional subcommittee. She remembers being in the Senate chamber when Idaho's Frank Church "made a speech written by the Wilderness Society about the wilderness, and he prevailed. The bill was passed."

Barbara stayed active in the wilderness movement for several years. She believed simply "that nature could do it better than anybody." She adds, "Primarily, we thought logging was a sinful thing to do. We were from the city and neither one of us had ever lived on a farm. We had grown up city people, and it is awfully easy for anybody that grows up in the city to have absolutely no concept of a renewable resource. You think of milk coming out of the bottle. It doesn't come out of a cow. You don't have to feed the cow. You don't have to shovel the poop, and all the rest of it. And so our notion was that this was a gorgeous wild area, and it should be kept that way."

But the wilderness movement came to frustrate Barbara, who begins to talk about her disenchantment as she gets up, goes to the kitchen, and brings fresh pears and homemade orange cookies to the table. We eat as we visit. Originally, she says, wilderness preservation was sought for places where "the hand of man is hardly seen nor felt." Successive wilderness bills, especially those related to the Upper Peninsula, became more inclusive, and Barbara began to see wilderness proposals involving land with railroad grades, roads, cart paths, mansions, boat houses, boat landings, and even a landfill—all land, she says, "with a history, with a human history of manipulation. . . . It was just so incongruous, I threw up my hands in horror. And I was one of 'em."

About this same time, the mid-1970s, Barbara went to a private landowners' conference. Part of the program was to visit the Alberta tract that marker Jim Johnson managed for so many years. Jim told me how pretty that managed forest was, and Barbara came away equally impressed. "They showed us different kinds of forest management," she remembers. "This one section had plots that every seven years they would cut in, and they had it all calculated—how much pulpwood and logs—and how every time, every cycle, the forest that was left was of higher quality."

Thus, Barbara ceased her support of wilderness areas and became a proponent of timber management. "What intrigued me," she explains with a softly closed fist, "was the idea of a piece of woods that would pay for itself, that was a

126

contributing part of the family system, much in the same way that a sheep dog is far more interesting to me than a lap dog. You know, it's a working dog, and it's a working forest that is producing something. . . . We became convinced that there was a better way than just letting nature blow the trees over every once in a while."

It struck her at the time that she and John could have their harvest and their crop. "We could have our woods and encourage our wildlife. We could manage along the creek where there's a lot of evergreens and wildlife, and on the steep hillsides for erosion, and not cut there. And the flatter areas, we could cut and take out the small trees and the overmature trees." So, they joined the Michigan Forest Association, had a marker come in to designate which trees to cut, and, while John taught electrical engineering, Barbara found herself locked into a ten-year contract to thin their timber. She learned to handle a chain saw and found herself girdling old poplar so they'd die. "If you cut them down, they'd wipe out a half an acre of young trees, so you girdle them and they die and they come down limb by limb." She made three miles of ski trails through the bush and found herself cutting down trees in the way, leveling off hummocks with a maddock, and filling in low spots with dirt and branches. Her labors were rewarded when the Michigan Forest Association officially dedicated their property as part of its statewide timber management program. In the small ceremony marking the occasion, a state forester used an increment bore to take plugs out of some maple trees; the growth rings showed double and triple the growth since Barbara's thinning. To see that result so quickly, "that was heartwarming, very satisfying," Barbara says.

Our conversation lapses as John stokes up the potbelly stove in the dining room. Then I ask Barbara to talk about the effect of land, not on her philosophy, but on her attitude about nature in general. She answers that the land is a hobby for her and John. "You don't like to think of the land as being your hobby, but that's what it really is," she says, "because we support the land, in a sense." Working the land has made her protective; when somebody criticizes her ski trail, she says, "it's like knocking my kid." And the work also has changed her view of nature. "It's an active force that you have to deal with all the time," she notes. "Sometimes it's nice, and sometimes it's pretty rotten. Nature isn't just a sweet mama with green hair." A little later she adds, "I look upon land much less romantically and much more realistically—not just the land, but the water and the rainfall and the snowfall and the wind and all of these forces that you can't do anything about. I get angry sometimes when they thwart me."

She explains. You fell a tree; it drops on your truck and you have to replace the hood and windshield: "My aim was off." You plant a garden; you dig up countless bracken fern roots: "horrible, big, woody ones." You scheme and plant a garden up in the field of the Rinkinen farm to get away from the ferns; rabbits and deer get the vegetables before you do: "You have to take what it [nature] gives, and it isn't worth it to a lot of people." You scheme some more and put a fence around the garden; a bear runs right through it and tears it down. You fix a cabin; you face getting supplies down the hill: "You just have to take it in stride, one way or another." Nature, she says, summing it all up, makes you feel "puny." She says that despite the challenges of living in the bush, the lessons are much different than, say, for pioneer families, who had to make nature yield or face death. She repeats again that to her and John, "it's a game."

Nevertheless, their many years of labor bind them to the land deeply, but, in a way, also futilely. "I find it kind of bothers me when I realize that it isn't obvious that any of our kids would appreciate inheriting this place," John says. "It would be a problem for almost any of them, we can tell at the moment, rather than something they would enjoy."

128

"None of them wants us to get rid of it particularly," Barbara adds, "but none of them wants us to keep it for them either."

"What makes it worthwhile?" I ask.

"I don't know," Barbara says.

"I don't, either," John adds. "I sometimes wonder."

"I wonder, too," Barbara says.

Then, except for the crackling of the wood fire, quiet fills the room for several seconds, and for a moment the silence seems as rueful as Swede Intermill's story about the lonely young woman in Misery country who years ago screamed into the night.

Doug Hamar

Pat Beyers

THE FACTORY

In the early days of hardwood logging in northern Michigan, lumberjacks floated logs down rivers to Lake Superior, where they were gathered into booms and towed to sawmills. This practice was common, for instance, in Misery River country where John and Barbara Clark have their cabin and mapling operation. The Ford Motor Co. was one firm that logged in that area, and its tugboats towed the booms northeast around Keweenaw Point and then southwest to L'Anse—a trip of more than one hundred miles. Log booms clogging the waters of Keweenaw Bay gave work to nimble gandy dancers with hobnail boots, thirty-foot-long pike poles, and lots of nerve. The men who did that work early in the twentieth century . . . well, says one old Keweenaw lumberman, "Our ranks are thinner."

Today truck drivers and twenty-four-wheel rigs roll out of the bush and haul logs to sawmills, and from there to wood product plants. In northern Michigan, that plant as often as not is Horner Flooring Co. in Dollar Bay.

This verse hangs on the wall in the office of Doug Hamar, president of Horner Flooring:

> *No disputing it—*
>
> *Not a doubt of it,*
>
> *No refuting it—*
>
> *No wear-out to it,*
>
> *Maple takes it all!*

Hard maple, that is, and from it Hamar's firm makes flooring that has sold around the world. Dollar Bay is not the likeliest site for a business that has received such recognition, for it is a modest town of eight or ten businesses and a few hundred residents. But Horner doesn't need population centers to thrive; it needs only timber. Situated in northeast Houghton County, the company is at the heart of Michigan's best hardwoods. The Keweenaw Peninsula doesn't produce enough hard maple to keep Horner's production line going, but the company buys 80 percent of its lumber within a fifty-mile radius of Dollar Bay, the bulk of that from Houghton and Keweenaw counties.

Ninety-five percent of Horner's product is hard maple flooring, most of it the traditional nail-down, strip variety. In addition to basketball gyms, the flooring is used for squash courts, racquetball courts, health and aerobic centers, textile mills, tobacco warehouses, multiple-use facilities, factories, post offices, and residences. All these uses aside, Horner is best known for its portable basketball floors, which one employee calls the company's "glory product." The portable floors grace gymnasiums from North America to Europe to Asia. Horner has supplied floors for the XXIII Olympiad in Los Angeles, nine NCAA Final Four men's college basketball championships, six NCAA women's Final Four tournaments, eight NBA All-Star Games, two Pan-American Games, the McDonald's NBA Open in Barcelona and Paris, the Seattle Kingdome, the Louisiana Superdome, Tokyo's Metropolitan Gymnasium, and the arenas of the Detroit Pistons, Milwaukee Bucks, Utah Jazz, Sacramento Kings, and San Antonio Spurs. The business that these portable floors brings in isn't that significant—Horner sells only twelve to eighteen a year—but Doug Hamar says that the publicity the company gets from them is invaluable: "You turn that TV on and watch that game and know the floor was made right here in Dollar Bay—you take pride in that. It certainly helps us."

The portable floors are popular for multi-use arenas that have many events, Hamar says. They take two weeks to build, two weeks to sand, paint, and finish, and generally just a few hours for a crew of five to install. The record: a crew of twelve assembled the Milwaukee Bucks floor in twenty-three minutes. The floors are comprised of 210 panels that are four feet wide; 196 are eight feet long; 2 are six feet long, and 12 are four feet long. They have metal joiners on the corners and fit together like a jigsaw puzzle. Horner has spent nearly thirty years developing the floors. In the 1960s, Hamar says, the panels were fourteen feet long and were awkward and cumbersome. In the late 1970s and early 1980s, the floor materials were refined to the product that is sold now. Most of the process was developed in-house, Hamar says—"an idea here, an idea there." The floors have absolutely no loose parts; they are shipped without so much as a single loose screw, nail, or bolt. They are popular, says one employee who works on them, because "they're easy to lay and easy to take apart." Horner has a patent on the design.

Horner sometimes rents the portable floors for one-time use at a special event. Then they are dismantled, shipped back to Dollar Bay, reassembled, sanded, refinished, and sold as used floors. When I visit Horner, five portable floors— two new ones and three rentals—are stored in a warehouse. One of the rentals was used for a NCAA women's Final Four tournament. I ask Hamar why he has that floor back and not the floor for the men's tournament. He says that a speculator bought the men's floor, cut it up in five-by-six-inch pieces, engraved them by laser with the names of the four final teams and the final score of the championship game, made 22,000 of them, and sold them for $24.95. Noting that the speculator claimed to have spent $75,000 on advertising, plus $55,000 to $60,000 on the floor, Hamar says that, from what he knows about it, the venture was "a moderate success."

Horner was founded in 1891 in Reed City, Michigan, but disaster hounded the firm. After a fire in Reed City, the company moved to Newberry, Michigan. After another fire there, it moved to Dollar Bay in the midst of the Depression in 1931. When the smoke cleared from each fire, there was opportunity: the chance, each time, for the company to move closer and closer to its resource, northern Michigan hard maple. John Hamar, retired president and Doug Hamar's father, bought the business from the Horner family in 1960—the year Doug was born. Doug Hamar has been president of Horner since 1990.

Horner's administrative offices are housed in a prefabricated home that is nondescript on the outside. Inside, the distinctive part of the structure is that most of the floors are hard maple. One, that of Lew Bosco, the sales manager, is all clear wood and thick with bird's-eye. It looks rich and gaudy.

During the two days I spend at Horner, Doug Hamar is the only person of the sixty who work for the firm whom I find wearing a necktie. He has worked for the company for twenty years, having started in high school mowing lawns and doing factory work. His father had two guidelines for him: a demanding one of making him start at the bottom, and a protective one of not letting him work with machinery lest he get hurt. Doug Hamar remembers begging his dad to do other jobs than some of the more menial ones in the factory, which he thought at the time were "an insult to your intelligence." Years later he can say with some detachment, "He gave me an appreciation for the many steps in milling maple flooring. I didn't realize it, but he was grooming me for the office."

Doug Hamar, who is thirty-six, has a bachelor's degree in mechanical engineering and a master's degree in business. Since he began working full-time at Horner in 1983, he's been in management, administration, finance, production, plus assistant to the president, vice president, and, as he puts it, "janitor today." He's proud of the firm that his father helped build. He attributes Horner's success to the fact that for one hundred years, the firm was managed by only two families. Still, he says as he points out the window at the factory across the street, "Quality starts with men in the plant. There're only certain things we can do on this side of the street. Everything happens over there."

The person in charge of the Horner product is Mark Young, son-in-law of John Hamar, who has worked for the firm for sixteen years. Today he is the vice president. His hiring caused no trifling problems with the rest of the employees, who resented the nepotism. "We knew it was going to be difficult, and it was," he says. Young, who is dark-haired and husky, has a disarming affability, however, and harbors no resentment for his initial reception at the plant. He speaks of the company's factory hands as good men and women and "the key to the business."

At age forty-two, Young exudes friendliness and self-assurance as he talks about hard maple's place in the U. S. flooring industry, which is dominated by oak. Eighty percent of the hardwood timber on the continent is a species of oak, he says, and it is a popular product for floors in residences. "It's a good thing people like it," he says, "because that's what's out there." Oak is an open-grained, dark wood; maple is a tight-grained, white wood. Young says the dark woods have dominated America's taste for many years, but white woods like hard maple and aspen are now in vogue, as they were before World War II. "The Scandinavian look—light blond woods—they're kind of popular again," he says.

Compared to the oak industry, however, hard maple is still an upstart. The oak industry produces nearly 200 million square feet of flooring annually; the hard maple industry produces about 10 percent of that, eighteen to twenty million square feet a year. Overall, maple comprises 3–5 percent of the total flooring market in the U. S. "We have our niche," Young says. "We do very well, but our industry doesn't dominate the wood flooring business by any stretch of the imagination." That niche is a cozy affair: all of the hard maple manufacturers in the U. S.—they number five,

135

including Horner—are located within a couple hundred miles of each other in Wisconsin and northern Michigan, and they constitute the entire membership of the Maple Flooring Manufacturers' Association. Young won't say what Horner's portion of that small market is. "We get our fair share," he says guardedly. "We know what it is, but nobody else knows."

Young is more direct when he tells me how maple flooring is produced from rough boards. The overview sounds simple: workers stack the boards, dry them in a kiln, take apart the stacks, cut the boards to width, plane them smooth and fashion a tongue and groove on the long side, cut out defects such as knots and cracks, put tongues and grooves in the butt ends, grade them for quality, and stack and bundle them.

Young takes me to the start of the process—a dock where truckloads of hard maple are unloaded. It takes a minimum of three semi-trailerloads a day to keep the plant's production line operating without interruption, except for a thirty-minute lunch break. A forklift brings a pile of lumber to the unloading dock, where two men with picaroons—pick-like instruments that have a slender point like a snipe's long beak—impale boards, drag them off the pile, and put them crossways on a conveyor. The picaroons work well in the summer; in the winter, when the boards are frozen together, sledge hammers are used to free them. The conveyor is part of a machine called a stacker, which sorts the boards in seven-foot rows, aligns the ends, and unloads them onto another pile. There, two more men put three or four hard maple spacers between the rows so that they are exposed on all sides to air. As the boards are stacked, hydraulic cylinders adjust the height of the pile according to its own changing weight, so that the men are always working at waist height.

The stacker is the first machine in a highly mechanized production line. When a big enough load is stacked, the wood is ready for the dry kiln. Young calls this the "green end" of the production line, where workers deal with green wood.

Jim Juopperi, forty-eight, who has worked at Horner for twenty-one years, is in charge of the stacker and dry kilns. I meet him on the first day of summer; the air is cold and the wind has a bite. Juopperi is tall and thin, with a bad eye, but he has his job well in focus: he must get most of the water out of the maple. There's plenty of it; 35–65 percent of the weight of green hard maple is water. In the summer, the water content will be on the low end of that range; in the spring, the water content will be on the high end. A board foot of hard maple weighs seven or eight pounds green and about four pounds dry. A board that is eight feet long and six inches wide will shrink a half-inch in the kiln. Most of the shrinkage is in the width, not the length.

To transform it from green to dry, the kiln regulates the temperature and humidity to dry the wood slowly. It takes eight to twelve days to dry hard maple, three to four weeks to treat oak. In the summer, the kiln temperature for hard maple is kept typically at 130 degrees Fahrenheit the first five days, 140 degrees the sixth and seventh day, and then increased ten degrees each day on the eighth, ninth, and tenth days. The temperature and the humidity in the kiln are critical, Juopperi says, and he tries to keep the humidity as high as he can without retarding the drying. "If it's too high, it ain't going to dry. It'll mold or whatever. If it's too low, it dries too fast, and that causes all kinds of problems." The kiln schedules and treatments differ among the different species of wood and at different times of the year.

Two aspects of the drying process are especially tricky— and constantly threaten to unravel Juopperi's calm, easy-going disposition: sticker stain and case-hardening.

Sticker stain results from the spacers, called stickers, that are put between each row of lumber to facilitate drying. They often leave a stain or shadow where they contact the lumber. Sticker stain "is the curse of hard maple," Mark

Young says. "It's God's wish." Oak, too, gets sticker stain, but oak is a dark wood and the shadow rarely is detectable. Hard maple, on the other hand, is white, so just a little stain shows clearly and becomes more pronounced when sealant is applied to the finished floor. The stain, Juopperi says, is a chemical reaction in the wood caused by heat, humidity, and sugar in the wood—"the same sugars that make maple syrup." He adds, "Maple can get stain just sitting outside. If it's real humid and damp and calm, we'll get it. There's not enough air flow." The best way to combat the stain is to get the wood stacked and spaced and into the kiln to control the heat, humidity, and dampness. "Then you can control your own conditions," Juopperi says. Sticker stain is largely a mystery that befuddles kiln driers. The stain doesn't affect production; it only affects grade. When sorting the wood by quality, the graders consider the stain a defect, like a knot or crack, and put it with other lower-grade wood.

Juopperi's second bugaboo is case-hardening, a condition directly related to the moisture content of wood. Wood absorbs water and gives off water in the same way that air does. Wood also is elastic, and it expands and shrinks depending on its moisture content. Juopperi tries to reduce the moisture content of maple to an industry standard of 6–8 percent. More water, and the flooring shrinks after it is installed and cracks; less water, and the wood becomes case-hardened. Juopperi says that kiln driers want more moisture in the sapwood than in the heartwood to keep the wood workable. If the sapwood and the heartwood—what Juopperi calls "the shell and the core"—have the same moisture content, tension builds up in the wood. "It's like a cramp in your muscle," Juopperi says. He combats this case-hardening by bringing the moisture content of the maple to 6 percent, then adding 1 or 2 percent moisture to take the tension out of the shell or sapwood, which absorbs the moisture more readily than the heartwood. Juopperi calls this "conditioning" the wood.

Every board's makeup is different, though, and case-hardened wood slips through the kilns despite Juopperi's ministrations. Case-hardened wood is harder to mill. Mike Carne, who operates Horner's planer, says that he can tell when case-hardened wood comes through the mill just by the higher whine of his machine. Mark Young can tell when case-hardened wood comes through the mill by a falloff of 5–10 percent in production. Normal production is seventy-five thousand lineal feet of lumber a day. To meet that need, Juopperi uses six kilns. He calls a load of lumber "a charge" and puts in about three charges a week. Thus, the six kilns operate, in effect, on a two-week rotation—three charges in every other week, three charges out on the weeks in between.

When the lumber comes out of the kiln, it goes to the production line. Three men unload the boards row by row, pile the stickers, and rip the lumber to the proper widths for flooring. After the lumber is ripped, conveyors take it to three workers who do what is called "defecting"—cutting out big flaws, mostly cracks and knots. This entire first phase of the milling is under the direction of Richard Laplander. He is thirty-eight, thin, balding, and solemn. The hardest part of his job, he says, is "just to keep going" for six days a week. He puts in five days milling the lumber and a half a day Saturday sharpening saws and repairing equipment. "Shit breaks," he says. Laplander has worked at Horner for twelve years. He says that he didn't want to go to college, adding, "There's a lot of wood in the area, so you end up in the wood work industry." He calls himself a "wood tick," explaining that, besides working with flooring all week, he has built a cabin out of ninety logs. "Wood is sometimes boring, but it's my life," he says. I comment to him about the tedium of the jobs. "You have to work yourself on a pattern," he says, himself a veteran of some of the jobs before he became a lead man. "You get into a motion and go with it."

After Laplander's crew rips and trims off defects, the lumber goes to Mike Carne's planer, called a "matcher," which works all four sides of a piece of wood. The matcher is as big as a small bulldozer and is the noisiest machine in the plant. Without earplugs, a person standing beside it for a day surely would incur deafness or madness. The matcher planes the top and bottom of the board smooth and puts the tongue on one side and the groove on the other. It also cuts two shallow grooves in the bottom of the boards; the grooves relieve stress and, by little increments, add up to less weight and cut shipping costs.

What bedevils Carne in his operation of the matcher is simply the hardness of the wood he mills. Hard maple is the hardest wood he works with; he says it is harder than red oak, birch, and cherry. Carne's planing knives also are dulled by dirt, case-hardness, nails, and bullets.

"Bullets?" I ask.

Some trees that are logged have bullets in them, he explains, representing "the ones that got away," by which he means deer. Carne, who is forty-nine, has been at Horner twenty-five years and has worked on the matcher for twenty-one. He used to collect the bullets as novelties, but gave up four years ago. He says that they show up once every two or three weeks. Bullets are a minor problem, he says, because the lead is soft. Nails are worse because they chip the planing knives.

Carne's years on the matcher include three years of training, seven as an assistant, and eleven as the lead man. His goal on the job: "Make sure the company gets what they want, when they want it." He says his machine doesn't have a temperament. "I've seen some machines that you'd think had a mind of their own," he says, "but after you learn the machine, you have tricks to compensate for that." He says the matcher, although thirty years old, operates at 80–85 percent efficiency.

After the lumber leaves the matcher, it has been nearly transformed, like a bud to a flower, into handsome flooring. It is of precise dimension, because Carne gauges the thickness and width down to a thousandth of an inch, and it is handsome. For commercial and residential use, the finished product is three-quarters of an inch thick; for industrial use, it is an inch or more thick. It generally has a two-and-a-quarter-inch wearing surface, with additional width for a quarter-inch tongue, but Horner also makes other flooring that ranges in width from half to double the norm. Horner will custom-make wood products, from teak boat decks to birch, oak, beech, and pine floors, paneling, and molding. Carne has at his work station a box containing about forty kinds of wood pieces that Horner has turned out for customers.

The next step in the production line trims out more defects and grades and stacks the flooring. Pete Alger, thirty-five, a fifteen-year veteran of the mill, is in charge of this stage. The lumber already has been partially defected, what he calls "rough-trimmed," and now his crew fine-tunes it. It ends up in lengths from one foot to eight feet. Alger's trimmers cut out knots, blemishes, and cracks and put tongues and grooves in the butt ends. Graders then sort out the four grades: bird's-eye ("What I hear is it's really helluva expensive stuff," Alger says); first grade (no knots and good light color); second-and-better grade (some color variation and tight knots); and third grade (dark wood, lots of color variation, open knots, and small cracks). The wood is then bundled and stamped for grade.

Floors are priced according to the grade of lumber in them. Installed, Mark Young says, a high-quality residential floor costs about fifty-five to sixty dollars a square yard—about the same as high-quality carpet. The biggest competition for hard maple floors is not residential carpeting, however, because residential floors are a small part of the business; rather, Horner competes with manufacturers of

141

synthetic floors for gyms and other large facilities, and the company's gymnasium floors range in cost from about fifty dollars to a hundred dollars a square yard.

The next stop for the flooring is the warehouse, and Alger takes me there. On the way he shows me his left index finger, or what remains of it, for he cut part of it off six months previously. He tried to free a piece of wood stuck in the chain drive of a conveyor system; instead he freed his finger of a couple of knuckles. "Stupidity," he tells me. "It sort of squished it off."

Alger introduces me to Gordy Wetelainen, warehouse lead man, then salutes, and walks off. Wetelainen, who is fifty, portly, and solicitous, has worked for Horner for twenty-five years and been a lead man for more than half of that. He stores more than maple flooring stock. His long buildings also house the material used for the sub-floor: foam, fiber board, metal locking devices, and sleepers—two-by-three-inch lengths of wood used like floor joists to support the floor. Wetelainen says the location of all the material is in his head; his partner calls him "a walking computer." He allows that some people think he's crabby, but explains, "I'm just kind of a perfectionist. I do what I do and do the best I can. I try to keep errors to a minimum. If I do make a mistake, it travels many thousands of miles. If it's the wrong material, it costs a lot of bucks." He ships all over the world.

One of the things Wetelainen likes about his job is meeting truck drivers from all over the country, and he is mindful of their needs. "They are losing money when loading," he says, "so I try to get them in and out of the shipping yard in an hour." He also tries to organize the loads that have more than one destination, what he calls a "drop," so that the truckers easily can cover the various orders with tarps. "You get multi-drop loads, you start scratching your head," he says. He remembers one load that had seven drops in three states. "That's my record," he says. "It's been quite a few years ago."

Wetelainen is proud of never having sent the wrong material to the wrong destination in ten years. Nevertheless, he says, "Sometimes it's kind of brain-wracking. Winter is really miserable." He explains that he loads trucks outside and the forklifts don't have cabs. "We dress accordingly," he says.

Inside the plant, the burr in the pants of the workers is not cold, but noise. On my second day at the mill, the production line is shut down for the installation of new equipment. But, when it and other machinery are operating, the long, narrow, cavernous building is filled with an incessant din that compels all workers to wear ear plugs while on the job. "That noise, the high pitch," says one worker, "you go home, you can't listen to the kids 'cause you hear the noise all day long. 'Just be quiet,' I tell them. That's why I go up the camp. I go up the camp three, four days at a time. Come back—I'm all nice and calm. Every once in awhile you gotta do that. I like the bush. I like to go out in the woods, listen to the birds and stuff. Nice, peaceful. Boy, the more quieter it is, the better I like it."

Speaking is Pat Beyers, who works on a crew of three that unstacks the kiln-dried boards, piles the stickers, and rips the rough lumber to width. I meet Beyers as he idles away the closing moments of his work day. He is leaning on a railing next to a large, thick pad of wet concrete that he has helped form and pour in the mill. "*Hyi, hyi*," he says. The word is an untranslatable Finnish expression for something that is unpleasant or revolting. Notwithstanding English and Irish blood on his father's side, Beyers is mostly Finnish and sprinkles his speech with Finnish expressions, often swear words. "What a day," he continues wearily. "When that concrete comes, you gotta go." The pad will support a new ripper. "At least it's supposed to be quieter than the old one," Beyers says.

When the whistle blows to end his shift, he and I go to a restaurant to talk about his job. Beyers, who is forty-two, is big-boned, and his middle shows signs of prosperity. His eyes are blue and they twinkle with a devilish humor that the tedium of the mill work hasn't dulled over the eighteen years that he's worked at Horner. He has a youthful face. A few vague frown lines cross his forehead, but deep laugh lines at the corners of his eyes overshadow them. His talk is sprinkled with frequent soft chuckles.

The first day I visited the plant, I was struck by the tedium of the production line. Workers stood at their tasks, going through well-rehearsed motions with a repetition of movement that would dull many people's senses. It was a grim monotony that was tiring even to watch—no single person important enough to stand out, but everybody essential; all tasks remarkable only in their ennui, but all of them absolutely necessary to produce from rough lumber a piece of flooring. So now, with Beyers, I ask whether he likes his work.

"Are you kidding?" he answers. "No way. No. It's boring all the time. And it's push, push. Oh, you gotta go, boy. Go, go. Production is number one. Can't even go to the can unless you find somebody to relieve you. And you get high-strung from all the noise all the time. It's surprising how much stress there can be. The hardest thing is the same routine every day. But everybody needs to work somewhere."

Beyers rotates among three jobs: gathering up and piling the stickers that separate rows of hardwood boards, feeding boards into the ripper, or handing boards to the person feeding the ripper. Beyers began working for Horner in the 1970s and quit after five years to go to school to be an auto mechanic and then to work in that trade. He made five dollars an hour—not enough for his family. "I liked my job then," he says. "I was a happier man. But I wasn't making a living." After a year, he returned to Horner, where the pay scale for hourly workers is between $7.95 an hour and $9.22 an hour, 50 percent more than he made as a mechanic.

Beyers has a wife and four children. He says of working at Horner: "If I was single, I wouldn't be here. But when you got kids, you gotta work. Simple as that." The responsibility of housing and feeding his family weighs on him, gives him those few frown lines across his forehead. He says that he has gone though a side of beef and two deer in nine months. He marvels at the appetites of his boys, who are in college and high school. He also has a grade-schooler. "Holy shit!" he says. "I hope the hell she don't grow up to eat like her brothers. *Oi saatana.*" *Oi* is an untranslatable expression of astonishment. Beyers' tone, however, isn't bitter, disillusioned, or profane. He swears out of habit, not out of hard feelings. He has an irrepressible good nature and is inured to the tough job he has.

"Do you dread going to work?" I ask.

"Not really," he answers. "If I'm home too long, I'd sooner be at work. It's not a bad job, you know."

Beyers even finds pride in his work. He says about Horner's famous portable floors and his own morale: "You get to see things like that in the paper. Like you see a ball floor you made, like for the Detroit Pistons, and it's all right, you know. Somebody comes up to you and says, 'Hey, you did a good job on this floor or that floor.'"

Beyers is active in the union for Horner's fifty-odd hourly workers. He has been the steward and now is the vice president. His wife Debbie, who also works, wishes that he didn't have those responsibilities because, she says, he worries too much about it. But his equanimity suits him for the job. "You gotta stick up for the men," he says, "and you gotta be fair with the company, too." He constantly teases Bill Dunstan, the supervisor of production, a salaried employee, and his boss. "I get on Billy a bit every day," Beyers says, chuckling. "I razz the shit out of him. Well, he's gotta do his job, but when I hear shit, he hears shit." With that touch of sass,

Beyers manages to defuse tensions and do his union work in-formally. The touchstone of his performance: the grievances that he *doesn't* have to file.

Joking with others punctuates Beyers's daily routine; it is to him like a cold glass of water on a hot day. He likes to put grease on a piece of flooring to tease the next person who will handle it. "Oh," he says, laughing softly, "you gotta bullshit around a little bit once in awhile. Otherwise it would be monotonous as hell." Watching him work with eight-foot-long boards, which will stretch to thirteen miles of lumber that he will handle in a day, I try futilely to imagine his daily drudgery.

"What do you think about?" I ask him.

"Hunting," he answers without hesitation.

He loves the bush, the hunt, and the camaraderie of camp life. And he accepts, with what to me seems like bravery, a job that forces him to dream throughout the entire year of two or three months of pleasure in the autumn—all to divert himself from a humdrum toil that would sap the spirit of a person with a soft mind. "My advice to anybody is get a good education," he says equably. "Don't get a job like this. But, you know, it's interesting—I made more money going back to Horner than I did as a mechanic. So, actually, I came ahead."

Mark Young is sympathetic with the humdrum lot of Beyers and his work mates. "This factory work—there's no way around it—there aren't any glory jobs here," he says. But, Young adds, the methodical routine turns out an ex-traordinary product. "All I've ever heard is that our flooring is the best and it better stay that way," he says. "That's what keeps our customers." He is a true-blue salesman, rhapsodiz-ing about Horner's product like a songbird serenading spring. "If you're ever going to sell anything," he says, "you have to love what you sell. And I love hard maple flooring. It's still an honest product. It's not a gimmick. We charge a fair price for it, but you know it's going to last a long time."

Takano Kimpara

THE INTERNATIONAL BUYER

Takano Kimpara, 55, of Shizuoka, Japan, is a wandering merchant, a man of logs and timber, of grain and figure, of curl and fine bird's-eye. Since 1968, he has searched the world over for beautiful wood, but his odyssey really began in his mind as a child. At age six, in 1947, he suffered a crushed leg and was confined to bed and home for three years. He passed the long hours and long days reading books about famous people and about people with handicaps. His own is a left leg that is three or four inches shorter than his right. When he stands upright and still, it dangles oddly in the air; it reaches the floor only when his right leg is bent noticeably. Kimpara says that his handicap is a friend, that his limp always helps people to remember him. Similarly, he considers as a blessing the three childhood years that he was on the mend: confined to his own house for so long, he dreamed of traveling afar. These days he does—to Australia for oak and eucalyptus, to central Africa for pure black ebony, to Indonesia for black-and-white ebony, to Myanmar for teak, to India for rosewood, to Finland for white birch. On and on he goes—"all over the land and country," an old cowboy I knew would say—including the U. S. for a dozen different woods, and northern Michigan for hard maple.

Kimpara is short and trim, with thinning black hair, broad shoulders, and a polite and accommodating manner. He is at once a man of the office and of the woods—a neat dresser with social grace, and a savvy lumberman with a hatchet that has a curved blade to take the bark off of trees, and a hammer head with die to put his family's stamp into the butts of the logs that he buys. "This is very dangerous," Kimpara says as he shows off the trademan's tool. "Tomahawk. My

149

nickname is 'Tomahawk.'" His favorite word in English appears to be *nickname*.

When I meet him the first time, in a motel in Baraga in the early winter of 1993, he apologizes for his English and, donning glasses, frequently consults a Japanese-English, English-Japanese pocket dictionary. When we go back and forth with the dictionary and zero in on a word or idea, he utters his most common expression: a long, enthused "ye-a-a-ah." Sometimes I don't know for sure his meaning. He says, for instance, that some people call him "left-handed" because of his bad leg. I take that to be a literal translation for *Lefty*. At other times his near-misses with English are sometimes more evocative than fluency. Thus, he says of a Keweenaw logger: "He is kindness."

In Japanese, Kimpara's name means *gold field*. It should mean *wood lot*, because some of his family has been in the timber business for generations. Kimpara himself, after studying business and law at a university, got a job in the pulp and paper industry. He didn't like the work, and he says now that he tried to "run away" from it, but he couldn't because that's where all the job opportunities popped up. Ultimately, though, his imagination created an escape. His father made softwood posts for building houses, and Kimpara says the operation was wasteful: 22 percent of a log was lost in the milling. Mulling the situation over, Kimpara began to produce housing material and veneer from the same log. Veneer—a thin layer of prized wood glued on a substrate of cheaper wood, often plywood—attains a look of richness and beauty that is more affordable than solid pieces of exotic or expensive woods. An acquaintance of mine once said of a skimpy piece of calf's liver served up in a Wyoming restaurant, "You could read a newspaper though it." One might say the same about veneer, especially Kimpara's. His is at least four times thinner—.2 to .17 millimeters thick instead of the U. S. standard of .8 millimeters, which means that Kimpara gets about 130 sheets of veneer from one-inch-thick material, compared to the U. S. standard of thirty-two sheets. Kimpara uses his veneer for an array of products—from furniture and paneling, to car dashboards and musical instruments. On our first visit, he shows me samples of his veneer. They are a little bigger than a business card. The lighter woods are dyed—two bird's-eye maple in purple and pink and a curly maple in raspberry. Two darker woods—Iowa walnut and Australian oak—are natural. Asked how many colors he dyes wood, Kimpara says simply, "Rainbow."

Kimpara says his business, which he started in 1967 at age twenty-six, is "not so big" but is known for products of exceptional quality. Kimpara says he has succeeded not through the help of others but by his own drive. He deals with 124 different woods around the world. He reminds me of an old man I once met in Colorado who liked to show off his walking stick. It was made of fifty-five different kinds of wood from every continent except Antarctica. Each piece was shaped like a cork—the bottom of one was the diameter of the top of the next—so the whole thing tapered uniformly, with a hole through the middle for a metal rod to hold the whole works together. It was a dandy cane, but to identify any of the woods, the man had to refer to a guide that he guarded with utmost diligence. Kimpara's chore is even more ticklish: he not only has to know one wood from among 123 others, but also how to tell the difference between the good and the bad in each species. That ability astounds me. I ask him how he can do that, and he searches for several minutes in his dictionary for the answer—talking in his native language with an assistant, consulting his small, worn dictionary, and then, with a tone of the discovery of the obvious, saying simply: "Perception."

Kimpara owns five veneer mills and one sawmill in his native land. He tells me that his main showroom exhibits 124 chairs. Each is handmade by the same craftsman, each is made out of one of his woods, each is worth forty-five hundred dollars. None of them is for sale. It is a splash of

151

flamboyance for someone with a subdued and humble manner, but evidently appearances are not deceiving because his firm, one of more than three hundred like it in Japan, has been routinely cited for excellence at Japanese trade shows. Some of his international clients tell Kimpara that he has the best veneer in the world. No bragging—just fact, it seems, because when Emperor Hirohito died, Kimpara was awarded the job of redoing Emperor Akihito's office in the royal palace. Kimpara used one wood, Japanese cherry, and says being awarded the job was an honor. His mother told him he should donate the material and the work. Instead, he ended up doing it for cost. He doesn't want to reveal the amount; a gesture so crass, I gather, would only hurt the emperor's "pride."

Custom orders are not uncommon for Kimpara. A more recent order is illustrative: a Japanese businessman wanted his entire house to be built out of one tree. The average Japanese home is about the size of a modest American ranch house, Kimpara says, and in California he found a tree big enough to build one house. The log he got from the tree was thirteen feet in diameter at the butt and more than one hundred feet long. It cost ten thousand dollars on the stump. Kimpara says buying, cutting, and shipping the tree was "very expensive, very, very."

How much did he make?

That was an old sale, he answers, and he only thinks of new ones.

He does say what kind of wood brings in the yen. He gestures at the furniture in his motel room—all made of composite material covered with plastic that looks like wood. Such stuff goes against the grain of Japanese, he says, who like not only real wood, but the best of it. They want wood with a small, tight grain that results from slow growth, instead of a big, loose grain that results from fast growth—a difference in appearance like that between finely layered Lake Superior agates and big-banded Brazilian agates.

Japanese also want their wood to have dense, smooth texture and generally very light, "constant" color. Lastly, they want a straight grain; if a wood has what Kimpara calls "cathedral" grain—oval patterns—the ovals must be small and complete circles; even then, they are inferior.

On our first visit, Kimpara shows me, besides samples of his veneer, his business card. Appropriately, it is made of wood—in this case, polonia, a Japanese ceremonial wood. Traditionally, Kimpara explains, a polonia seed is planted at one's birth, the tree is cut down at one's marriage, and, because it absorbs moisture and is resistant to fire, the lumber is used to build a wardrobe for expensive kimonos. Listed on the back of the business card are some of the places where he does business these days: Italy, Germany, France, India, Malaysia, Myanmar, Singapore, Thailand, Taiwan, Hong Kong, Russia, and the U. S. He is on the road each year for about six months. During that time, he makes about fifteen buying trips. Five or six of them are to the U. S. In northern Michigan he buys mostly maple, but also a small amount of cherry, red oak, basswood, yellow birch, and curly aspen. U. S. woods comprise only about 10 percent of Kimpara's timber purchases. The biggest by volume: white oak.

Kimpara first came to the U. S. in 1983, when he was forced to look farther afield because some countries in Southeast Asia where he did a lot of business had established export bans. On that first trip, he traveled by car from North Carolina to southern California, staying on the same latitude as Tokyo. He bought only yellow poplar for furniture and woodwork, but he says it was a fast-growing, big-grained wood, not white enough, and therefore "not so interesting" to the Japanese market.

A year later he came to the northern U. S., seeking bird's-eye maple. The bird's-eye pattern was not new to him; a prized wood in Japan, zelcova, has the bird's-eye figure, and, judging by Kimpara's description of it, the eyes are as big as silver dollars. But bird's-eye in hard maple was new to him, and an engaging set of circumstances stirred his interest. First of all, he met with a Wisconsin trade delegation that was ballyhooing that state's wood, including bird's-eye. Secondly, a Japanese airline company, Ana, was building a big and luxurious hotel at the Tokyo airport and had put out orders for bird's-eye. Thirdly, Kimpara took in an international veneer trade show in Cologne, Germany, and got his first good look at bird's-eye. Lastly, when Kimpara headed to New York via London and waited for connections at Heathrow Airport, he started chatting with the fellow next to him. It was Prince Andrew. "What kind of woodwork is in your mother's office?" Kimpara remembers asking.

"Bird's-eye maple," he was told.

Like a hound on a trail, then, Kimpara landed in New York and struck out across the country, staying north, looking around for wood in general and for bird's-eye in particular. It was the first of four successive, but not successful, trips. On the first one, he bought a small amount of bird's-eye in New York and lost money. On the second trip, he did the same and lost more money. On the third trip, people at Ohio's Mead Paper headquarters steered him north to Michigan's U. P., where he bought bird's-eye near Iron Mountain, on the Wisconsin border; he lost money yet again. The three trips in three years had three things in common—bird's-eye maple that came with a gray tinge, with dark mineral deposits, and with red ink. Finally, on his fourth trip, Kimpara probed farther north into the U. P.— from Baraga north into the Keweenaw, and there he found his mark—"very pure white" bird's-eye, with small hearts, tight texture, and little mineral stain. He began making money on the species. Now he buys all he can get on and around the Keweenaw. "Bird's eye country," he says, adding: "Nickname."

Kimpara generally likes the Keweenaw's long winters, but he gets a taste of an unusual one on his four trips during the winter of 1993–94: an unusually mild December with the first bad storm on Christmas Day; a lulu of a January when temperatures down to twenty-five degrees below zero make every footfall squeal like a wounded rabbit; an early March thaw that is interrupted only briefly by what locals say is an annual tradition—a St. Patrick's Day storm; and, lastly, an early spring, drab no matter when it comes, with black humps of road sand looking just like slag piles along the roads. On his last trip in early April, Kimpara says that he is disappointed in the winter of 1993–94. It is too short, he says, so that he comes to the U. P. only four times instead of his usual five or six. The longer the winter, the better for him, the colder the winter, the better for him; for he comes to the Keweenaw only in the winter, when the sap is out of the trees and doesn't discolor the wood. During the Keweenaw's summer, when Kimpara is not in Baraga or Houghton or other points north, he is likely to be roaming the same latitude in the Southern Hemisphere. He says the climate along the Forty-fifth Parallel is good for people to live in and for trees to grow in. He allows, though, that this preference is "only my philosophy." He also likes his trees cut during a waning moon and an ebb tide—all of which, he explains, means that trees are in a condition "just like sleep" and the wood cells are dry and shrunken, what Kimpara calls "very small wood" with "good texture." Here again he allows that this practice is "only my sense," but he adds that he's met a "timberjack" in Panama with the same belief.

In his search for good bird's-eye, Kimpara singles out Paul Clisch, a Baraga logger. He stumbled across Clisch through word of mouth after he got lost while poking around the U. P. in the mid-1980s. Kimpara says Clisch is knowledgeable and honest and coached him about the figured wood: How the sign of exceptional bird's-eye is occasionally on the butt of the log, where light-colored rays sometimes radiate from the heart outward like bicycle spokes. How the bird's-eye showing up in the butt can be misleading, so that it is advisable to cut a three- or four-inch-thick piece—a "cookie"—

off of the butt of the log to get a second look at it. How a good way to test for mineral stain is to inject water into the butt of a log with a syringe—the mineral deposit will often betray itself by discoloring the wood.

Largely because of Clisch, Kimpara is an old hand now with bird's-eye, which is in great demand in Japan. Kimpara says he and other wood merchants "created" the big market for bird's-eye in Japan. One of his own contributions to the promotion was a ploy so simple it is remarkable. Thus, bird's-eye veneer of "not so good quality" covers the front desktop of the luxurious Ana Hotel, while bird's-eye veneer of exceptional quality covers the doors of the main rest-rooms in the lobby. Kimpara explains that only one of a party of guests signs in at the desk, while everybody who's anybody passes through the restroom doors; thus the strate-gic placement of the best wood. In any event, he says that a woodworker in Hiroshima tells him that the bird's-eye figure is popular—"very easily sold out"—because it resembles "ex-cellent marblestone."

The most maple Kimpara ever bought was on a trip in 1990: 170,000 board feet, a fair share of it bird's-eye. On his December 1993 trip, he buys only 10,000 board feet, a small amount of it bird's-eye. He has to put a handsome price on maple to make a profit. He pays $1,500 to $2,000 for plain maple veneer, routinely ten or twenty times that for bird's-eye. Added to the cost of purchase, he pays about $8,500 to ship the logs from the southern shores of Lake Superior to the western shores of the Pacific, in railroad containers that hold about 3,500 board feet. Taking no chances with sap stain, Kimpara buys logs in the winter and ships them in refrigerated containers—to Seattle by rail, then to Japan by ship. Dockside in Japan, he pays an import fee of $1,400 per container. Then—in a demonstration that it's a small world, and that if it's buggy and muddy in the northern Michigan bush, they know about it in the dockside warehouses of Shizuoka, Japan—-he pays about $800 to spray the logs with insecticide, and $1,200 to wash the logs to get rid of insect larvae and dirt, the insects and their eggs being "a very big problem." Kimpara then stores the logs for about a month— at a cost of more than $6,000 a month for each container— in refrigerated warehouses, right alongside the tuna fish. Then he ships it to his mills. By the time the wood gets there, where he can dry, mill, and market it, he has invested, on the average, $18,000 for each container, or about $5,000 per thousand board feet of logs. Then he dries it, mills it, and markets it. The long and expensive journey has been reflected in the cost of the finished product. In 1992, at the height of the maple market, a sheet of the least expensive plain maple veneer, one meter wide, one meter long, and 0.17 millimeters thick, cost about four dollars, while a sheet of excellent bird's-eye veneer cost eighty dollars. In 1995, those respective prices were more than halved and ranged from a dollar to thirty dollars.

Up until his visit during the winter of 1993–94, the most Kimpara ever paid for bird's-eye was at a rate of $50,000 per thousand board feet. It is not the most expensive wood he deals with. Equally or more valuable are Japan's zelcova, which Kimpara says is like elm, Indian rosewood, and English maple with the fiddleback figure, which occurs throughout the wood in soft maple, but only one to two inches deep in hard maple.

"Why the difference?" I ask Kimpara.

"Only Christ knows," says this Buddhist.

When I first meet Kimpara in mid-December of 1993, he says the bird's-eye market in Japan has peaked and started to decline. What's more, he says, the word among timber merchants in Japan is that virtually all of northern Michigan's good bird's-eye has been cut. A month later, in mid-January of 1994, on his second trip of the winter to the Keweenaw, Kimpara calls me to say he feels poorly from high blood pressure and might not be able to see me. But, on his last night, he calls and we meet at a lounge. Paul Clisch, says

157

Kimpara, has "found out" a nice area and has shown him a beautiful bunch of bird's-eye. Kimpara says a few of the trees have eyes as big as peanuts. He is utterly effusive: "Very excellent bird's-eye . . . Very happy . . . Very successful trip . . . No more sick." He says that he and Clisch aroused the bush. They clapped, laughed, hugged and hooted, and they shook hands. Crows reveled overhead. "We are happy, and birds, too," Kimpara says. "Mister Paul Clisch knows everything. He is 'Mister Bird's-eye.'" And then he adds: "Nickname."

As Kimpara and I talk, he smokes Mild Seven cigarettes, a Japanese brand, drinks two bottles of Bell's Amber Ale, and munches on strips of dried kelp, which are about the size of pieces of bacon, but paper-thin, and black and shiny. He offers me a piece. I try it and make a face. "Mister Paul Clisch" didn't like it either, he says.

I don't see Kimpara on his third trip of the winter to the Keweenaw, in early March of 1994, because he has to make a quick side trip to northwestern Wisconsin to buy oak. But I learn later that, on that trip, he bought more than a hundred logs, much of it bird's-eye, from Clisch. He also bid on forty-two logs from Lake Superior Land Co. and got thirty-eight, the other four going to an Italian buyer. Then, on our last visit in April, at the start of spring breakup, Kimpara says he bought four more logs from Clisch, and twenty-six from Lake Superior Land. Five of those twenty-six were exceptional, he says, and he paid the most he's ever paid for bird's-eye. How much? Kimpara doesn't want to say, so his competitors won't know how much he bid. But I talk with Michael Lorence of Northern Hardwoods about the prices for bird's-eye. The industry range, he says, is ten dollars to seventy dollars a board foot. Lake Superior Land's Keith Brey says that he's seen bird's-eye fetch as much as eighty-five dollars a foot.

Kimpara says that his bird's-eye, all veneer, will go to three hotels of the Prince hotel chain. He has purchased enough in northern Michigan for hotels in Tokyo, Sapporo, and Hiroshima. There are four more to be built and he must supply the bird's-eye. He figures the prime wood in the northern Michigan market is virtually exhausted for now—thus the exorbitant prices—so next winter he will shop for figured wood in Maine, other parts of New England, and upstate New York—settling, he says, for wood that he expects to be inferior. He will also shop for bird's-eye in Russia. He hasn't been there since 1983, but will return in the fall of 1994. He's got a hunch there's bird's-eye in Siberia. "I think—not yet sure," he says. "Only my feeling."

Meanwhile, the plain maple market will stay strong. The reasons: Japan is enduring a recession and inflation, and people are looking for cheaper products; also, the government is starting to build 1.5 million houses and will need cheaper material.

Kimpara says that there is much waste in American forests; that too many trees are standing. He calls America "a young country and ground." By *ground*, he means soil, and he says that it is generally much shallower in America than in Japan, and, as a result, most of the trees, especially those in the Midwest, are smaller. Japanese soil is deep and rich, he says, and the trees generally are much bigger. Regardless, he likes to do business in the U. S., where transactions are simple and straightforward. He gives me a sampling of his experience elsewhere:

Canada: the dealing is easy, as in the U. S., but he has stopped going there because of competition from big companies.

China: the timber is fine, the bureaucracy is slow. Up until 1987, he went there every month. Now there are too many buyers to go that frequently.

Russia: like China, fine timber; like China, slow and bureaucratic. He doesn't like the pressure for, and prevalence of, under-the-table deals.

India: some years ago, he and the government of India built a veneer mill that employed more than five hundred workers; but Indians, Kimpara says, are "not so good work-ers." When he visited the country, they worked; when he left, they went on strike. He lost money, he says, and so did the government.

Many other countries: often there is pressure to buy inferior wood. Sellers keep lowering the price trying to make a sale. Kimpara says about refusing to buy inferior wood in the U. S.: a no is a no, and the decision is accepted.

Of all his travels, trips to the U. P. are Kimpara's favorite, or, as he says, "best for me." He likes the people and says the wood is "very special"—in what way, though, is unclear, for he deals in beauty, but when I ask him what his favorite tree is, he says simply and without hesitation: "The profit kind." Yet he has his eye on more than just the ledger book. He says that he is lucky to have friends like Paul Clisch wher-ever he buys. "Our hearts are very tight," he says of Clisch. I ask him to explain; he says that the two of them share "vir-tue," and Clisch is "like family." He says that he has one such friend in each state where he buys timber, including an Amish man in Pennsylvania, from whom he buys yellow poplar, white oak, and black cherry. On trips there, Kimpara stays in the man's house, where there is no television, no ra-dio, no electricity, no power tools—instead, candles, crosscut saws, and a horse and buggy. There is nobody who lives that way in Japan, Kimpara says, and he admires the Amish way of living as "very good and natural."

The Amish man told Kimpara the legend of the dogwood tree, and Kimpara recounts it for me: Christ was crucified on dogwood, and ever since the tree is small instead of big. Also, ever since, it has red spots, representing Christ's blood, on its white flowers.

"Do you believe that?" I ask Kimpara.

"I believe, I believe it," says this Buddhist.

It is not unusual for him to link spiritualism to trees. Trees are "near" to people, like animals, he says. He proffers his business card again, which carries the family symbol, representing sun and spring water, which are important for both trees and people, he says. His family has had that logo since one forebear, an uncle more than one hundred years removed, adopted it. The uncle is noted in Japan and has been written up in schoolbooks because of his way of life, called a *doo*, which emphasized honesty and simplicity. This uncle finished a successful and storied life by establishing six two-thousand-acre tree plantations and a foundation for silvicultural research, all of which the government still manages today. Another long-gone relative, Kimpara says, helped him and Clisch find the good bird's-eye stand in the winter of 1993–94. In a dream, the old man's ghost told him to not give up looking for it, to keep trying hard to find it. He says the dream was sent by his grandfather. When I ask him to explain the dream, Kimpara searches and searches in his dictionary for the right word, and his finger eventually stops at a Japanese word for "oracle" or "divine message."

Even a more mundane item, Kimpara's business card, is vested with higher meaning. The back of it carries his company's slogan, "Wood is good." Kimpara explains that the saying is a pitch to the people of his country, who are, he says, overly tense and busy. The slogan is meant to suggest that wood is natural and holy—and something to give people pause. In effect, it suggests that his countrymen slow down and smell the koyamaki—a wood with a sweet scent, "like sassafras"—that is commonly used in Japan in bathrooms (Kimpara has four in his house; they are unlike in that each has a different dominant wood; they are alike in that each has some koyamaki). The message—advice, really—is ironic, though, for Kimpara's life is anything but relaxed. He is not well. His doctor tells him that he suffers from high blood pressure, that his visits to the cold climes of northern Michigan are bad for him, that he is a victim of too much stress. Indeed, he tires of his travel regimen. He tells me on one trip: "Very tired. Yesterday am sleepy in quiet hotel. Some fever." I ask Kimpara why his work is so hard on him. He explains with one fact: he has had to sell high-priced bird's-eye for firewood because some mineral stain sneaked by his heedful eye. "Dangerous business," he says. "Gamble."

When not on the road, he does get some respite from his pressured life. He slows down in his small garden, where, along with plants and turtles and goldfish, he and his mother, who is eighty-five, tend to a bonsai plant, a two-hundred-year-old azalea that he bought for nine thousand dollars. In his garden, he says, there is no stress.

Kimpara has been at his business for twenty-seven years—one year more than half of his life. In that time, he says, he has cut down thousands and thousands of trees, so that his traveling has turned from a dream into a debt. He says he is now in the autumn—what he calls "the fall"—of his years, and he will try to square up with the natural world that he has lived off of. He says that flowers, grass, and trees are fixed in place but yield seeds for the future, while he has wandered the world over and has sown no seeds, even though he has fathered four children, which he says is an entirely different matter. To help nature along, he will establish tree plantations in Japan. He figures it is his responsibility to the next generation. He and a Japanese forester already have begun experimenting on 143 acres. He hopes eventually to plant ten thousand acres, just like his forebear did more than a century ago. Not yet aware that nobody knows how to grow bird's-eye, Kimpara says he recently told Clisch that he would like to grow the figure on his plantation. "Paul Clisch is laughing," he says drolly. Nevertheless, he says of the plantations: "This is my dream."

They are a part of his own *doo*, which Kimpara, after turning over many leaves of his dictionary, describes as *naturalism*. Nature is God's way, he says, and this man whose search for fine timber has put him in touch with royalty and commoner alike, says, "Only God can make a tree and a human, I think."

Paul Clisch

THE DAYS TO COME

Part I

"You should be able to log it forever."

A verse from a folk song, sung by Odetta, says you can always spot a lumberjack because he stirs his coffee with his thumb. The general drift of the rest of the song is that a lumberjack is a crude and unusual fellow with rocks for muscles—and for brains, too. Paul Clisch would see himself in that song. "I'm just a dumb logger," he says when comparing himself to college-educated foresters. But his self-effacing manner belies his sawdust-and-sweat sheepskin, which allows him to talk about logging and timber management with an authority as sure as hindsight. What he has to say makes him somewhat of an anomaly—a logger acknowledged as "superior," who generally recoils from overcutting, and who says of the northern hardwood forest: "It's gotta be managed. We're not doing a good job of it now."

I learn about Clisch and his reputation for the first time quite by chance, when I meet Glen Chrisinske, of Munith, Michigan, at the Indian casino in Baraga. We sit together at a gaming table, start small talk, and, oddly and quickly, our conversation skips from good wagers to good timber. Chrisinske says his land has just been logged, and, laying down a five-dollar blackjack bet, he tells me about a sure bet—a logger named Clisch, who, Chrisinske says, knows more about maple in general, and bird's-eye in particular, than anybody else, anytime, anywhere.

So it is that one summer evening I find myself in Clisch's home, sipping the boiled coffee people made before percolators and eating zucchini bread made by his wife Helen, who comes from a settlement near Baraga called Section Twelve and who helped Clisch with bush work in the early days of their relationship. I ask Clisch why Chrisinske sings his praises in four-part harmony.

The answer: Chrisinske owns 120 acres of land in Skanee country east of L'Anse. It seems a forester had marked Chrisinske's timber and advised him that his cedar was especially valuable. Chrisinske, who Clisch says is a "tough dealer," then put his timber on the market, but nobody bid on it because of the rough terrain, what Clisch calls "tough ground." Chrisinske then asked Clisch to look it over. Clisch roamed the acreage to prepare his own bid. He remembers telling Chrisinske: "Your value is in the bird's-eye and birch. Your cedar isn't worth the powder to blow it to hell." Clisch wound up logging the stand and took out between eight thousand and ten thousand board feet of bird's-eye—more than two truckloads—and paid Chrisinske more than double the minimum bid. "So that's why I think Glen and I get along pretty good yet," Clisch surmises in a towering understatement. The whole deal hinged on bird's-eye's cryptic features—and Clisch's ability to spot them where others can't. It's simple, he often says: "If it's got dollar signs, it's bird's-eye. If there's no dollar signs, it's plain maple."

Paul Clisch, who was born in 1940, is a Santa Claus with muscle. Broad-shouldered, big-boned, big-bellied, he has a good humor that matches his heft, laughing often, usually at himself. Clisch started in woods work as a lad during and after high school in the 1950s, when he peeled bark from pulp sticks for his father. His one tool was a spud—a leaf spring from a car, sharpened on the leading edge to get under the bark, and taped on the trailing edge to make a handle. Clisch and his wife peeled a hundred cords of pulp one summer, earning the down payment on their home. "We both know what spuds are," Clisch says. Such work hinges on weather and season, so when Clisch married, he figured he needed steadier work so he took a job in a welding shop. He lasted two years. "I couldn't take the inside part," he says with a shudder even yet. He then went back into the woods, back out into the weather, and felt as content as an acorn. Clisch worked for other loggers for ten years, had a partner from 1973 to 1983, and has been on his own since then. Now, he says, "I work by myself for myself. . . . If I don't work, nobody works."

His labors these days are especially fruitful, Clisch says, because of the market for maple: "It's just unbelievable." In the past, the market followed a steady up-and-down pattern from year to year. "I guarantee," he says, "if you stayed up there one year, down you were going" the next. In more recent years, up to about the early 1990s, the price of maple held steady. Then the demand for maple and the prices for it went up. Between 1992 and 1994, prices changed five times —each time rising, all totaled doubling the price: up to $400 for 1,000 board feet of plain maple; up to $2,000 or more for 1,000 board feet of veneer; and up to the heavens for bird's-eye. Even the value of the poor grades went up. "Five or six years ago, you had to beg these sawmills to take your logs," Clisch says. "Six years, eh?" These days, he can tell the same buyers to come out to his job in the bush and bid for his logs. He likes that luxury. "I always sell to whoever, you know, I feel like it."

Clisch is ever on the watch for figured timber, and he has gotten Takano Kimpara interested in not just bird's-eye but curly maple, too. "I told him, 'The curl—that's nicer than bird's-eye,'" Clisch remembers. Kimpara bought a few logs; then, for a while, took all he could get. Even though Kimpara's interest "toned down" by 1994, curly maple is "a good item," Clisch says, sounding a bit like a five-and-dime clerk. Curly maple is more common than bird's-eye, Clisch says, because bird's-eye occurs virtually exclusively in hard maple.

In thirty-five years of woods work, Clisch says, he has seen but two soft maple with bird's-eye in it. Curly maple, on the other hand, occurs in both hard maple and soft. In recent years, Clisch also has sold to Kimpara one quilted maple—the stuff Fred Aho found in Pennsylvania ("It's a blister on the log, is what it is—the bigger the blister, the better"). All figured maple isn't valuable. Clisch has to keep an eye peeled for a figure that only he has told me about: thumbnail maple, which, on the stump, shows up as little depressions in the wood of a size that is obvious by the name. The depressions are actually miniature seams with ingrown "bark pockets," and Clisch is fortunate if he can sell a thumbnail log for plain maple.

Spotting figured timber while it's standing is not always easy, Clisch says, echoing others. To find any figure, you first have to have a clue, and even then, you can be fooled, Clisch says. "You can look at a bird's-eye tree and actually have eyes in the bark. Okay. But you can take that [bark] off and instead of having any bird's-eye, it could be curl. It could be curl and bird's-eye. It could have thumbnail. It could have thumbnail and bird's-eye. There's five or six combinations. You look at that tree on the outside, and . . . you'd swear that's bird's-eye, and it isn't. It might be bird's-eye. It might not even be nothing there." Sometimes curly maple shows up in the bark like bird's-eye; sometimes Clisch can spot it where a tree breaks off at the notch when he fells it; most of the time Clisch finds it like all the figures—simply by pounding lots of bark off a log with the blunt end of a hatchet. "It's tricky business," Clisch says. "You gotta know your timber."

He does. "I got a pretty good eye," he allows, adding that he acquired it simply by cutting down trees and "paying attention." In his eye, then, there is no substitute for plain old woods work. Sometimes, Clisch says, markers with only book knowledge can miss bird's-eye completely or "see" it where it isn't. Clisch recently looked over a timber stand that, said a forester to the landowner, contained a lot of bird's-eye. "I couldn't find hardly a *decent* bird's-eye," Clisch says. It's a matter of importance (false expectations can breed animosity between owner and logger), and it's a matter of reputation ("So who's going to be the turkey? It's going to be the logger—'He swiped the bird's-eye'"). So Clisch says he will not cut for anybody who doesn't trust him. "You don't trust me, then I'm not going to log your timber," he says, "because your timber doesn't mean that much to me." The price he won't pay for any timber is his good name. He wants one word uttered when people speak of him and his work: "fair."

One measure of Clisch's fairness is the standard contract for his work, which stipulates that 55 percent of the payoff from a cut goes to the landowner and 45 percent to himself. "If the price goes up, the landowner gets more money and I get more money," Clisch says. "The price goes down, the landowner gets less and I get less. So this kind of a deal is probably the best deal in the whole works." When a stand has lots of veneer or figure, Clisch gives the landowner an even better split. In such situations, he says, "I can take less and still make more." Some loggers don't pay a fair price, he says, and only self-policing separates the honest from the dishonest. He adds that most landowners, like Chrisinske, don't know the value of their own timber, wouldn't know bird's-eye maple from garden seed, and can be easy marks. Clisch says a logger can cheat a landowner by lying about the quality of trees he cuts and about the number of trees he cuts—also by overcutting a stand. "It's sad," Clisch says. "A lot of 'em have been taken advantage of." For his part, he says simply, "You get people to trust you. You have to in this business. You can screw these people so bad."

He treasures his good reputation. Many landowners simply tell him, "Use your own judgment" or "Cut it like your own." That faith might be wayward, though, for Clisch doesn't even trust himself enough to mark his own land; as a hedge against opening the canopy too much, he says he'd be too conservative with cutting. "I'd leave trees where I shouldn't be leaving them," he says. Generally, then, he expects the landowner, including himself, to entrust the job to a good marker.

I ask Clisch if I can accompany him for a day of work. He says yes—on a job west of Baraga, but first he must finish a job south of Baraga, on some hilly, rocky land near Nestoria. We make plans to meet in a week. Then Clisch warns me that I won't like what I see. I ask why not. He says he's cutting the stand too heavy. I ask why. He says that's what the owner wants, and there's no convincing him otherwise.

I get up to leave, and Clisch follows me down a long hall and out the side door. It is near dark. "I've gotta put music on for the deer," he says. I'm taken aback. He notices, then explains that the music keeps the deer out of his vegetable garden. They were coming every night—wouldn't even wait until dark. And they were fussy—would pass up the carrots and cabbage to get to the broccoli. Clisch first tied up his dog by the garden, but the deer simply went to the other end. Then he sprinkled Irish Spring soap around the perimeter of the garden, but the deer were dauntless. Finally he put speakers outside and tuned in to an all-night radio station. That worked like a song.

A week later, I call Clisch to confirm our rendezvous the next day.

"Should I bring a lunch?" I ask.

"You're not getting mine," he says.

I reach his home just outside Baraga before dawn. Clisch is limping. He hurt his foot the day before on that rough going near Nestoria. "Geez!" he says, "that was a tough day yesterday."

The pre-dawn air is cold, and a fog hovers over the land

like bad spirits, but Clisch is chipper. "I never have a bad day," he says of woods work. "I'll just have one day better than another."

I tell him that I've read that logging is one of the most dangerous occupations. "I heard that, too," he says. "I don't know." In thirty-five years, he has had one bad accident and one bad predicament. The accident occurred when his saw kicked back and cut his forearm to the bone. "Hell's bells," Clisch says, remembering the ride out of the bush with one hand on the steering wheel and the other hand gripping the hand on the steering wheel to stem the flow of blood. His chain saw had a kick-back safety feature, but he had taken it off because in the winter, when he wears choppers, there's not enough room to get a good grip on the saw. Clisch's predicament occurred several years ago when he was unhooking a tow cable that ran from his skidder to a stringer, the long stem that he tows to a landing and that he will buck into shorter saw logs. He had the skidder idling in neutral without the emergency brake on, and it slipped a bit and pinched his leg between the stringer and the machine. The accident occurred in mid-afternoon. While he watched his leg swell to seven inches larger than its normal size, Clisch sat on the stringer and waited for his wife.

"She knew something was up when you didn't come home?" I ask.

"When I miss supper," he says, "there's problems."

Clisch is logging just a few miles west of Baraga, and on the way from his home to the job site, I ask him why he took this job of overcutting. To him, the matter is not dirty work, rather just work, plain and simple. He says that if he didn't cut the stand, somebody else would; that if it's got to be done, he may as well do it. It's not only doing the inevitable, though. Clisch also believes that he cuts more carefully than others and does less damage to the forest. "Whatever's left," he tells me, is "not all beat up." One reason: his one piece of heavy equipment, a skidder, at eight feet wide and twelve feet long, is medium-sized. He doesn't even dream about big feller-bunchers that grab hold of a tree and cut it down, and skidders that are much wider than his and "take out" more small trees when operating. About his equipment, Clisch says simply, "I'm practically back in the horse days."

I also ask Clisch if winter logging is especially tough.

"No, no, no, uh-uh," he says. "That's some of our best logging."

He explains that in the summer, when roads are wet, it's sometimes hard to find timber to cut. In the winter, he uses his skidder to keep such roads cleared of snow so they freeze and are passable. The cold, then, is good for logging, and it is bracing, too. Clisch dresses in the wintertime "so that you gotta work to be warm," and he tells his wife: "I don't want a thermometer that works. What you don't know won't hurt you."

Clisch and I arrive in the bush early and sit in his pickup truck to await daylight. Soon we glimpse patches of horizon through the trees where the sky lightens and begins to reach overhead and penetrate the forest. Outside, a soft dripping sounds on the forest floor. I am puzzled.

"Rain?" I ask.

"Dew," Clisch explains.

When it is light enough, Clisch wastes no time in transforming the woods from repose to labor. He takes some aspirin for his tender foot, gets out of the truck, and tops off the bar oil and gas for his chain saw. In the winter, he'll do that the night before, but nature is not as exacting in summer, so Clisch works at a more leisurely pace. When he starts his saw, gone is the sound of the dew, and gone is the earthy scent of leaves and ferns. Instead, the smell of gas and oil fills the air, smoke envelops Clisch like an aura, and the saw putt-putts like a purring kitten.

Clisch has five tools for his work: two ten-horsepower Husqvarna chain saws with twenty-four-inch bars—one for felling clean-barked trees, and the other for cutting chain-

172

dulling, muddy stringers into logs; one yellow plastic wedge, which he carries in his hip pocket, for coaxing trees to fall this way and that; his skidder for dragging logs from the bush to the landing, where he stacks them for loading on a truck that he will hire; a maple stick, eight feet, six inches long and about as big around as a broom handle, for measuring the logs that he cuts out of stringers. "We don't carry no axe," he says, and he uses his wedge sparingly. "Dad would shoot me. They wedged everything. Now you just let 'em fly."

His irreverence is mere fancy, for he takes great care in deciding where he'll "throw" a tree. Between watching him all day and talking to him all day, I get the gist of his work pattern. He assesses the location of trees, the lay of the land around them, and tries to fell all of a group so that they point generally in the same direction—parallel to a small wash, for instance, so he can drive his skidder up the wash, hook on to, say, four stringers, and thereby minimize the damage that the skidder does to the surroundings. Before he begins cutting, Clish also takes note of other standing trees that he absolutely doesn't want to damage. And he checks for dead branches or leaning trees that might come down on him like wrath from above. Once he zeroes in on a tree, he considers its natural lean, for he must dump it in that general direction; but he can, if he wants, aim it elsewhere. He sets his sights first by where he cuts his notch. It is a wedge cut out of the tree trunk; it looks like a gaping mouth; it functions like the pride that goeth before the fall. The notch cut, Clisch then goes to the other side of the tree and begins his cut. Partway through, he might use his yellow wedge to relieve binding or coax the tree a little this way or a little that way. And sometimes he uses what he calls a "hinge." Picturing the cross section of a tree as the face of a watch, the notch goes from eleven o'clock to one o'clock; that is the general direction the tree will fall. Clish might cut the tree from, say, one o'clock clockwise to eight o'clock, but not from nine o'clock to ten o'clock. That uncut part will function like a "hinge" or pivot point when the tree starts to lean over, and it will help swing the tree left. One other consideration that helps Clisch aim a tree is the extra weight of leaves and branches. The sun draws them to one side of the tree or other ("Just how it pulls flowers and everything else," Clisch says), and he'll often use that tendency for help in dropping a tree where he wants.

With all that savvy under his belt, and like an old hand at any toil, he makes it all look easy, of course. But, in other respects, this day is not easy at all, for Clisch limps from tree to tree. "It helps when you got a better foot," he says.

He makes his cut on a tree where it begins to flare out as the beginnings of roots—usually knee-high or lower. When he cuts, thick streams of sawdust fly outward, dirtying him and marking the spot like a grave blanket. Clisch's saw sounds just right: it neither screams from underwork nor labors from overwork.

When a tree falls, it is a cacophonous affair—first cracking at its base, then cracking louder as it topples and falls, breaking branches off of its neighbors, or breaking in two the neighbors themselves, and then hitting the ground with a thud. Once it's down, neighboring trees continue to wave and swish as they shake off the assault. It is like the sky falling, and those tremulous neighbors quake a while before steadying up.

Many are the dangers of felling trees. "You gotta be careful or you're a dead duck," Clisch says. A falling tree can hit a dead tree, snap it off, and send it wildly who knows where. A falling tree can get hung up in another tree. That happens early this day, and Clisch fells another one across the leaner to knock it down. This solution complicates the skidding, but it works. The perils of wind are obvious for loggers. It and rain and spring breakup keep them by the home fires. "You try to shy away from trouble, is what you do," Clisch says.

When the tree starts its downfall, Clisch gets out of the way, but today he is in no particular hurry. He usually moves as fast as he can, but he can't this day because his foot hurts too much. "Damn that foot, I tell you," he says. Indeed, after cutting awhile, "Ouch" cries his foot, and he pauses for a spell. "Usually we don't take a five," he says, "but that foot wants a five." On a regular, steady day, he can tell time by how long a tank of gas lasts—about an hour and a half. Despite the dangers and occasional hurts, he likes woods work. "Yah, I tell you, it's a nice place to make a living." He never dreads getting up in the morning, and he likes working alone.

Once prone, a tree can still be dangerous: a branch on the underside can be bent and, when cut, unleash itself like a mad snake. When the trees are down, limbed, and topped, safe and sound, Clisch drives his skidder into position— plowing over saplings in his way. He maneuvers so the back of the skidder and its towing cable are near four stringers. He circles each stringer with a choker cable that he attaches to the towing cable, then he bounces down the bush road, like a traveler with tin cans tied to his tail. Clisch always skids as he cuts, particularly in the winter when he could lose logs under new snowfall. The skidder cost seventy-five thousand dollars. Clisch has to cut enough timber to pay himself, maintain the skidder (which has four big tires that cost almost a thousand dollars each, are five feet in diameter and have deep treads that get cut by rocks), and pay it off. "If you don't make it," he says, "you've got nobody to blame but yourself, put it that way." He always manages his skidder payment, but he balks at buying and using his own truck, which would increase his overhead and his worry.

After he reaches his landing, Clisch unhooks his four chokers, maneuvers the stringers in order with the small blade on the skidder, then moves the skidder away. He bucks the stringers into logs, tucking his measuring stick under one arm while he cuts. It looks awkward, but he works with ease. He cuts from the top and from the bottom—even straight into the middle with the tip of the saw, a ticklish, even dangerous maneuver. Four logs thus disposed of, Clisch takes another break. "Damn that foot, I tell you."

Before getting into the truck cab, he shows me his axe. The handle is broken, the bit dull. "In the old days," he says, "if you had an axe like that, they'd shoot you. Your chances of getting bruised are better than getting cut."

We get in the cab and talk again about the worth of work versus study. "The problem with some foresters," he says, "these guys don't cut a tree down. They're not learning, you know. That's the whole thing right in a nutshell." What they have to learn, say, about spotting bird's-eye, isn't in books, Clisch says. "I never went to school. I never studied a book. But you pay attention to what's going on and after awhile it's just common sense. I mean, you don't even have to be very bright."

Years ago, Clisch says, woodsmen like his father, working with crosscut saws and horses, went through the land and learned more about timber than he will ever know; by the same token, Clisch says that he has "zillions" of hours on a power saw and has learned much more than the average forester. "When you cut it and skid it, you learn things," he says. ". . . I'm not bragging or nothin', but I probably got more knowledge on bird's-eye than I'd say ninety to ninety-five percent of the people, maybe even more."

Clisch says bird's-eye occurs where maple is dense and grows especially slowly. For instance, he says, opening up the canopy too much can cause a tree to grow too fast and actually leave the bird's-eye behind, buried in old wood, missing in new wood. He thinks soil is a factor in the occurrence of bird's-eye, but not location—say, a hilltop or a gully. He says Fred Aho's "corset" shape occurs—but only on a small percentage of bird's-eye. He also says that the chances of finding bird's-eye increases in rough, rocky ground and in the company of hemlock. Beyond that, he stumbles trying to explain the telltale features. "You can look at a tree, it looks like bird's-eye," he says. "I'm not saying I'm right all the time, but there's something about a bird's-eye . . . I can't explain it, but there's something different about it."

The color of the bark?

No.

Something about the leaf?

No.

What, then?

"I don't know. It just stands out. I don't know why."

I can't decide whether he's being cagey or candid. But, whatever the whys and wherefores, loggers need the bird's-eye. "You bank on bird's-eye being there," Clisch says, adding, however, "It's always a gamble." The bet is increasingly a long shot. "We'll never get to the end of it," he says, "but there's less of it. . . . Everyone went after the bird's-eye when prices were high," so much so that in some areas crews with helicopters ranged through the bush just looking for the figure and cutting it. As of 1994, Clisch says, "There's not much that hasn't been cut."

The situation is a repeat of the past. There was a big, big demand for bird's-eye in the 1960s, Clisch says. For a long time after that, the supply was low and the demand, too. For the past several years now, the demand has been strong again; Clisch says that in the late 1980s and early 1990s, bird's-eye buyers came out of the woodwork. Bird's-eye, he says, was "hot," high-priced, and drew big money to the peninsula. Later, in another interview at his home, he will use Kimpara as an example: "He's not flying back and forth here five, six times a year because he's broke."

Clisch revels in some of his experiences with bird's-eye buyers. "Years ago," he says, "they seen eyes, they bought it, you know." So he talks of knocking bark off of logs "where you knew there was eyes, just about"—eyes that perhaps weren't in the rest of the log. During the early part of the late-1980s, early-1990s market, the sellers had more savvy than the buyers, he says. "The buyers came over more green than most of the guys that sawed it. Now that's gone, too. They got smart." He laughs roguishly, but he jests; if he doesn't jest, he exaggerates, because he has taught more than just Takano Kimpara about good bird's-eye, about whom he says, "I don't like to see him get taken."

While the demand for plain maple remains strong, Clisch believes that the market for bird's-eye has peaked once again and will level off for awhile or even go down. "It's a fad, really," he says. But a mystery, too. "It's unique. Face it. They can't grow it."

Like Fred Aho did, Clisch saves his bird's-eye for a fall or winter cut to avoid sap stain, what Clisch calls "a grey-type deal." The stain "ruins it for just about anything," he says. He has found one bird's-eye on this cut near Baraga. He knocked a small piece of bark off. He says, "I know it's got one eye anyway." He says he'll show it to me. Then he concludes his break with, "Well, I suppose . . ."

Clisch works till noon. Then once again we sit in his truck and talk about timber. Trees, he says, are a natural, renewable resource. He contrasts the forests with the Keweenaw's copper mines: "A mine hits copper—you're never gonna get another piece of copper out of that mine. But trees are always going to come back. . . . I told this one person, I said, 'God put these trees on the earth to use.' If he wanted us to look at them, they'd stay there and nothin' move—but they keep on reproducing themselves and everything else."

Clisch doesn't understand the environmentalist who would severely limit or curb logging and about whom he says, "They don't use no common sense at all." On the ladder of intelligence, he puts them on the lowest rung; on the ladder of irritations, he puts them on the highest rung. They flatly exercise this good-natured man. They are a black cloud in his life always, the cause of a black mood this day. "I start when I want, and I quit when I want," Clisch says of his work. "It sounds good, but it isn't." The reason for his disillusionment: environmentalists. They are why he won't encourage his sons to become loggers. "Too many people telling you what to do and how to do it," he says. He talks about some virgin white pine in the extreme north of the Keweenaw. Some environmentalists want to protect that stand from cutting or, the way Clisch looks at it, let it die, fall, and rot for "ground feed." Says Clisch: "Who the hell's gonna build a house out of a bunch of rotten wood? As a matter of fact," he adds with the proverbial tongue dipped in acid, "I'd like to be able to get a forty right next to one of those environmentalists and clearcut. I'd make the biggest mess you ever seen in your life. That'd be my most satisfaction I could ever get out of life, if I could do that."

"Really?"

"Yah."

"But you're one yourself. You're one who tends to undercut."

"I still would do it."

Clisch's rancor, though, is not in tune with the day, and it soon tempers to an appreciation of warm sunshine, a languid breeze that passes through his truck cab, a cozy lunch—and no bugs. "You wonder why I work in the woods," he says with a sudden calm. "It doesn't get any nicer than this. We get about three or four days like this a year."

He has sandwiches, an apple, and tomatoes. I have venison sausage and crackers. He shares the tomatoes, I share the venison. Clisch says: "My dad always said, 'A picnic every day.'"

Clisch saves the seeds from his apple to suck on in the afternoon. "That's my tobacco," he says.

The agreeable interlude ends with the words, "Well, I suppose . . ."

179

We leave the truck and walk up a grade to the lone bird's-eye tree he's found so far on this cut. Clisch has an expectant air; much money could be involved; he has made as much as seven thousand dollars on two bird's-eye logs.

I ask him how big the eyes get.

A quarter of an inch, he says: "Biggest I've ever seen and not too often."

We reach the bird's-eye. Dots bigger than BBs show in the bark. Clisch pounds off a section of bark a foot square, going through the gray outer bark, going through the brown bark beneath that, going through the tan layer beneath that, all the way down to the slippery, white cambium, revealing, incredibly, the telltale marks of not only bird's-eye, but also curly maple and thumbnail maple—all three, but in just that spot. He takes more bark off; the rest of the tree is plain. "That'll never sell," Clisch says. "It doesn't have enough eyes to see." We leave. "Now I'm mad," he says. "Shouldn't have showed you that tree. Have to go back and work now. No money for that tree."

Clisch takes the skidder back to where he is cutting a quarter of a mile away. Once again he hangs a tree up and knocks it down by hitting it crossways with another. He also leaves a barber chair on one cut, long pieces of sapwood standing on the stump like a clump of palm branches at an altar. "My dad would have conniptions," he says of the barberchair. "He'd fire me as a son." I tell him that what he's doing doesn't seem like an overcut. He says it's because of his bad foot; he's only taking the easy stuff. I go back to the truck and take a nap in the warm cab. Soon, the sound of the skidder heaving down the road to the landing wakes me up. Clisch gets out to cut the stringers into saw logs. "Geez, that foot!" he says.

I stroll to the part of the stand where Clisch has already cut. The terrain is made up of gentle folds of land with small, dry drainages accenting the undulations. With foliage it would look pleasing, but it is a tangle of tops and limbs and stumps. The logger who cut it has a complex ethic about his work.

Part of the ethic respects the landowner's right to do what he wants with his timber. Another part lets him overcut, should he choose, just for the money. Here Clisch's attitude about logging is mirrored by his feelings about some of his own land. On one eighty-acre parcel, he has a big, deep, spring-fed pond that is stocked with trout. Fishing would be good, but herons and otter feed on the fish, and Clisch says of them, tolerantly, as he might about any logger forced to do a slash, "They gotta eat, too, I guess."

Beyond these working rules, Clisch also is reverent about the trees that give him his livelihood. "I always cut them down," he says, "but I'm never able to stand 'em back up." And he recognizes his debt to both the past and the future. "Someone has to leave the ones before you." The extremes of his thinking, then, range from the lofty to the practical: save trees for the future if you can; cut it too heavy if you must. Overall, though, his views center on this firm conviction about northern Michigan's maple: "You should be able to log it forever—if it's managed. But, there it is, it's gotta be managed. . . . I always come back to managed."

Carl Puuri

THE DAYS TO COME

Part II

"You're sitting on real gold—green gold, but it's out there."

On a cold day in September of 1994, a lowering sky hunkers over the land outside L'Anse. There is snow in the air and flimsy patches of it on the lee side of piles of logs scattered about a wood yard. Twenty people, all members of the Michigan Forest Association, gather just inside the trees skirting the staging area, away from the bite of the wind. Ferns and leaves are starting to turn yellow, the first hue of autumn; but the cold overwhelms the prettiness.

All but one person are bundled up in winter clothes: chukes and mittens, scarves and earmuffs, and heavy coats. One old woman uses a ski pole as a walking stick. One man, from Alabama I learn later, wears only a trench coat—no hat, no gloves.

"Aren't you cold?" I ask.

"Yessir," he answers.

Small talk dwindles and dies as Dick Bolen, chairman of the Western Upper Peninsula Forest Improvement District (FID), begins to talk. Bolen is pitching the advantages of membership in the co-op that he oversees. The FID helps members inventory, mark, cut, and market their timber. All members' logs end up at one of the district's two wood yards, one in Gogebic County to the south and this one in Baraga County. Log buyers like the convenience; the members have the advantage of strength through bigger numbers and bigger volumes than any single seller would have. "We're trying to get top prices for the wood," Bolen says. It is the co-op's

183

job to keep track of where a log came from, who cut it, who bought it, and where it's going. "The bookkeeping gets kind of interesting," Bolen says. "We haven't lost a log."

As Bolen speaks, a snow squall whips across the yard.

"Want to walk a little further into the bush?" Bolen asks.

"About two miles," someone says.

Despite the cold, Bolen quickly warms up to his subject, about which he feels passionately. The FID, he says, has more than seven hundred members who own collectively about 150,000 acres of land in the western U. P. "We're now large enough so people talk to us," he says. "We're making money for our members." He says that more and more people tell him the FID is doing a good job.

What Bolen and his foresters are doing is managing land through selective cutting. "Do the job right and build your value," he tells his listeners about timber management for the long haul. "You gotta stick with it." The temptations to cut too heavily are especially alluring because of an unprecedented bull market for timber. "We've never had a market like now," Bolen says.

As Bolen speaks, the wind dies down, the sun shines through some breaks in the clouds, and the little patches of snow shimmer brightly. Shortly, the group leaves the shelter of the woods and walks among the piles of logs in the yard. The district is trying to sell as many logs for veneer as possible, and FID forester Jack Hornick talks about veneer and veneer buyers with a blend of instruction and wit.

He says that the demand for white birch—what he calls "paper birch"—red oak, yellow birch, and sugar maple is so good that the size of the small end of veneer logs of those species has been lowered a few inches. He says that small hearts generally are desirable in maple, but one buyer wants more red heartwood for maple door panels. He says that a straight seam on one face of a log, which at one time knocked it out of the veneer class, is acceptable now because veneer plants also slice logs instead of just peeling them. He says that one veneer buyer with a hernia doesn't roll logs over to check the face on the ground. He says that one buyer who *always* rolls logs over to check the fourth face didn't do so on his most recent trip. Hornick is at a loss to explain why. "I don't know if he caught two fish on the way up or what." Hornick says the cold sometimes makes buyers inattentive and hurried. "The weather works for you once in a while. . . . It's most interesting what happens across the year."

Halfway through Hornick's wood yard tour, clouds beat back the sun and once again the day is grey and cold. Alabama, who, I learn, is the absentee owner of eighty acres in Rabbit Bay, far to the northeast, looks chastened.

Gathered in the wood yard is plenty of hard maple and several other species of wood: soft maple, yellow birch, basswood, oak, and black ash. Hornick comments specifically on the black ash, noting that some Japanese timber merchants buy it because it is similar to polonia, the ceremonial tree that Takano Kimpara described, which, Hornick says, is getting rare. "You never know where your breaks are going to come from," he says. He pauses, then asks, "Any other questions? —No, I can't tell you how to keep warm." The group disperses to cars; some will rendezvous at a coffee shop. I make my way to where Alabama is getting into his car. Remarking on the affairs of the morning, he says: "I gotta get me some education."

Subsequent to the wood yard briefing, I meet with Bolen at the FID headquarters in Hancock. He is articulate and commands the respect of the foresters employed by the FID: "All the success we've had is because of Dick Bolen," one says. Bolen explains that the FID originally covered six western U. P. counties: Keweenaw, Houghton, Baraga, Ontonagon, Gogebic, and Iron. In 1995, the district expanded its boundaries to include Marquette and Dickinson counties.

The FID's 758 members include private, non-industrial landowners, some local government agencies, and absentee landowners from all fifty states. Northern Michigan's was the first timber owners' cooperative in the U. S. It remains the only timber co-op of its size, duration, and breadth of service, Bolen says. A few others in southern Michigan and other states are simply "brokers" of timber, he says.

The fundamental goal of the FID, besides getting good sales of members' timber, is to create what Bolen calls "a working, thrifty, growing forest" that will create jobs and improve the local economy. "We can help promulgate the forest and make it a quality forest, if man assists a little bit," he says.

Armed with that conviction, Bolen scorns what he calls "charlatan" loggers who overcut for the big, short-term dollar and leave behind a forest lot that might take 50–100 years to rebound. Neither does Bolen understand the person who opposes any logging at all. "Big, old trees," he says, "you watch 'em die, but you want to keep 'em, so I don't know— there's people like that." In Bolen's ideal forest, then, no big, old timber falls and rots, no poor timber thrives. "The quality of the forest will improve, logging the way we log," Bolen says. His guiding philosophy, like that of so many others in his profession, is selective cutting. His tool is the chain saw. Together, the two shape a forest that grows not only saw logs but also sawbucks. "There's one heck of a lot of value in owning land," Bolen says of northern Michigan timber. "You're sitting on real gold—green gold, but it's out there."

Cashing in on it is the reason for the existence of the FID. The co-op balances what's good for the forest now and what will be good for it down the trail; perhaps a thinning now and a veneer cut later; perhaps one hand in the hip pocket for immediate cash, and another hand in the vest pocket for long-term investment—overall, improving each owner's site and at the same time improving the entire forest.

"Landowner helping landowner" is the informal motto for the FID. Says Bolen: "It's probably a socialist kind of approach, you know, but it works." In one sense, it is ironic that the FID started out as a government program and was located in the western U. P., where timber men are farthest from the Midwest's markets and where many Finns reside who historically have had a proletarian strain that shies from big government. Of the FID, Bolen says simply: "They stuck it here—for better or for worse, they stuck it here—and it's working."

Events leading up to the FID weren't the happenstance that Bolen suggests. The project was the direct result of Michigan politics, an arena where U. P. legislators once had power incongruous to their numbers. Russell "Rusty" Hellman, a retired, twenty-year veteran of the Michigan legislature and one of the men behind the formation of the FID, recalls events leading up to the establishment of the co-op.

Hellman, who was born in 1917, is a northern Michigan native who has had an unwavering interest in the wood industry. "I was born back of a sawdust pile," the Dollar Bay native says, "I lived back of a sawdust pile, I'm probably going to die back of a sawdust pile. So I've been interested in lumber and forestry and so forth all my life."

That interest proved remarkable, because Hellman's work as a state legislator from the U. P. was remarkable, too. It came during a period of Michigan politics when Hellman and two other U. P. legislators were called "the U. P. mafia." Why? Two reasons, both based on geography. One—the essentially rural U. P. legislative districts were more stable than their urban counterparts downstate, which resulted in long tenure. Two—legislators from lower Michigan cities went home every night, a convenience that U. P. legislators were a far ride from enjoying, which in turn resulted in night work on committees and choice committee assignments. "That's where the action is, that's where the power is," Hellman says now. It was a time of unabashed pork barrel politics; over twenty years, Hellman accumulated thirty-seven ground-breaking shovels from projects in his district. His partiality to the U. P. is displayed by a trademark item: a tie clasp that he designed, made of copper, in the outline of the state of Michigan, but with the Upper Peninsula three or four times bigger than the Lower Peninsula. He passes them out like some people do advice.

Hellman brought his political suasion to bear on his abiding interest: the Michigan Department of Natural Resources. And of that agency's work, Hellman was most interested in the state's forests. That self-imposed charge was evident the first year he was elected, 1960, when he began making speeches to his colleagues about Michigan's forests. His message was straightforward: "When the auto industry leaves Michigan, all we will have left is our forest, and we better start taking care of it." As the chairman of the DNR oversight committee, Hellman controlled the agency's budget in the House and took the opportunity to put his money where his concern was. This lawmaker, who in the 1930s worked for ten cents an hour at Horner Flooring Co., included in the DNR budget $1 million a year for sixteen consecutive years—1964–80—for forest improvement.

Do falling trees in a forest make a noise if nobody is there to hear them? From his office in Lansing, Hellman heard them all—and he was distressed, in part because the timber trade's name for good hardwood forestry, selective cutting, became a blasphemy. He says: "We supposedly had selective logging. We did all right. We selected the best tree and the good tree and cut that down and let the junk grow." So, this son of Finnish immigrants organized a trip of state officials, citizens, and foresters to Finland, which, he says, has a splendid reputation for timber management. Finland's forests, Hellman says, are in "tremendously good shape." The trip by the Michigan contingent was marked by two main occurrences: a look-see at one of northern Europe's biggest timber cooperatives, and a meeting with foresters from the consulting firm of Jaakko Poyry, Inc., which, at Hellman's instigation, was later hired by the state to assess northern Michigan's timber.

The study was completed in the mid-1970s. Three consulting foresters reported that, without major changes, "the prospects for forestry and the forest industry on the Upper Peninsula are poor." Says Hellman: "They were very knowledgeable and they were also very disappointed in our Michigan forests. They said there was nothing but junk."

The Jaakko Poyry report, which focused on the forest as primarily an economic—not recreational or aesthetic—resource, is a rap sheet of misdeeds. It said that northern Michigan's forests were poor, its forest utilization was poor, and its forest industry was poor—a predicament resulting from, at worst, greed; at best, ignorance or perhaps just shortsightedness. Of the timber land studied—a wide swath of land that ran from Lake Superior to near the Wisconsin border—only 5 percent could be considered well-managed, the Finns said. The rest was like the withered vine.

The forest industry could thrive in northern Michigan, Jaakko Poyry said, because the world consumption of wood is growing and forests are more valuable; because the western U. P.'s climate, terrain, and site fertility "are good or even excellent for growth and for harvest"; because labor costs are moderate; because the network of roads, towns, and public services is good; because terrain makes logging conditions generally easy; and because the region needs new industry.

The assessment of the forests themselves was dismal. The report said simply that there was too much harvest of high-quality trees, and too little harvest of low-quality trees. The result: there were too many trees per acre, too many overly small trees per acre and too many poor quality trees per acre.

The Finnish consultants noted that growing and harvesting of wood is incredibly expensive, and the tendency is to cut heavily the trees worth good money and be done with it. They said that practice is folly, and they outlined alternatives: educate non-industrial private landowners about the benefits of good management and of cooperation; give those same landowners financial help to mark, thin, plant, appraise and sell their timber; entice to the area industries that would create markets for forest products—including veneer mills for high-quality timber and pulp mills for poor-quality timber.

The Jaakko Poyry study was not uniformly well received, but it had one enduring offshoot: the founding of the Forest Improvement District. Jaakko Poyry recommended cooperation among timber owners, and Hellman took the advice to Lansing, where he stuck $2 million into the state budget to form and fund the cooperative for five years.

Ten years later, Dick Bolen, like a breeze rustling the leaves, speaks quietly but emphatically of the co-op and its efforts. Translated into logging, that message means selective cutting of the western U. P.'s maple-birch stands, to which, he says, there are some hurdles to give pause.

The first is a bad attitude. "Americans," Bolen says, "value their cars more than their forty acres."

The second is the delusion that northern Michigan's timber resource is inexhaustible, even in the path of heavy cutting. "If we're doing the right work on the landscape, we'll have it forever," he says. "If we don't, it could be easily gone just like the white pine is gone."

The third is the greed of some loggers. "Possibly," this mild-mannered leader says, "using a crude term, they're mercenary. . . . They clean the forest to fatten their wallet, and they have no further interest in the future of this part of the country. We," he adds of the FID, "have an interest forever."

The fourth hurdle is the begrudging acceptance that any new kid on the block faces, a problem compounded because the new kid was spending government money. In the FID's case, any of its foresters on anybody's forty raised the eyebrows of neighbors. Bolen says, though, that the FID defended itself by telling those people, "We're not unfair competition, just competition."

The fifth, last, and most peculiar hurdle: the false security of a handout—in this case, the very grant that created the co-op. "Quite frankly," Bolen says, "maybe the best thing that ever happened was for our state dollars to get dried up." The state's annual grant of $400,000, he says, created an atmosphere where "there's not a whole lot of incentive to get off your butt and maybe do some of the things you ought to be doing." The five years of state money didn't mean the FID foresters "were sitting around being lazy," he asserts, but they "weren't looking seriously enough at how to be self-sufficient."

The grant allowed the co-op to mark and market timber for nothing, a service that became a quandary. "Never give anything away free," Bolen says now, "because people just don't respect something free." The district had to discard the soft, alluring mantle of public money and don the hair shirt of the open market, he says. And once the co-op began charging for its services, he adds, the reaction on the part of the landowner was, "I'm willing to pay the freight." The freight was $17.50 an hour, for foresters' services, and a percentage of the timber sale.

Bolen predicts profits for the district by the end of 1996. He also sees continued growth. By 1994, the growth had occurred gradually, but, Bolen says, "All of a sudden we've created a presence" in the marketplace and "people are starting to look at us." More presence begets more notice, more notice begets more presence, until good woodlots make good neighbors. "As people began to say, 'What's happening on my neighbor's land?' we just grew and grew," Bolen says. "We are going to be an important part of the forestry community in Michigan," Bolen says. Where will it end? Bolen dreams of a co-op working with 750,000 acres in the U. P., and he dreams of the FID having its own veneer plant. "Opportunities get broader as the district gets bigger," he says.

The whole matter is not just about land and trees, he concludes. Rather, it's about "people that have grabbed hold of this dream, if you will, and are going to make it happen." For, what it all boils down to, Bolen says, is not only good silviculture, not only good economics, but also a principle of stewardship of nature that "is absolutely right." Speaking is someone with an eye quite far down the bush road, for Bolen feels a responsibility to "those kids who are going to be here in 2010, and their kids."

If Bolen ponders the needs of future generations, Carl Puuri, the head forester at the FID before he retired in late 1995, regards equally seriously the trees themselves. "They're living things," he says, "and if you treat living things right, they'll treat you right. You treat the forest right, it's going to produce a lot of income, it's going to produce a lot of pleasure, it's going to produce a lot of wildlife, and it's going to be a beautiful place."

In describing all that bounty, foresters, like Swede Intermill, are always talking about carrots. Puuri does, too, and more so. He says: "If you have a carrot patch—kept taking the very best carrots, with each weeding you'd just have nothing but runts left. If you're growing pigs, if you killed off all the best stock—ate all the best stock—before long you'd have a whole herd of runty pigs. . . . And the same principle goes with forest management—you retain the best growing stock with each harvest, you would have the best seed stock to produce the next generation of trees. That's the principle."

As tidy as all that sounds, Puuri admits those words can be like so many whispers in the wilderness. "It's easy to talk a good game," he says, "but the proof of the pudding is going out and seeing what's being done." So we hop in his pickup truck and bounce and slide along dirt roads that are alternately high-and-dry washboard or low and soupy—all of them marked here and there with bullet-dented road signs. We drive for three hours as Puuri shows me well-managed and poorly managed timber stands in north-central Houghton County.

He has a wiry body, a gravelly voice, and a slow way of talking that forestalls any unnecessary fuss. He began a career with the U. S. Forest Service in 1960, at jobs in Washington, D. C., Oregon, Washington state, and southern Michigan. He retired in 1986, when he moved back to the Keweenaw. Puuri likes working in the bush and helping trees grow. "It's more fun than playing golf," he says. No matter where he is, in the rain forests of the Northwest, or in the hardwood forests of northern Michigan, where he was born in 1933, he is content. To him, the woods is a chapel. "I'm just totally at peace being in the forest," he says on this autumn day, when the trees on high and the seedlings underfoot begin to look like an artist's palette. The sight is impressive, and it stirs Puuri. "You can't help wondering about trees and tree buds and things like that," he says, adding, "The forest grows, and that's the fun of it. It's really a dynamic place." The idea of a tall oak growing from a "lowly acorn" is the stuff of meditation for him.

The roads we poke along are between the farming area of Puuri's boyhood home near Liminga and Atlantic Mine and the Salmon Trout River, which is east of the land that Swede Intermill logged. Where Puuri picked potatoes in the 1940s and 1950s, forty-foot-tall red pine grows today. "That's how fast you can grow trees if you set your mind to grow," he avows. "It's kind of nice to be able to shape the forest around your old home."

A forest should not be like a machine shop where the squeaky wheel gets the oil; rather, in the forest, the problem tree should get the axe. But some stands that Puuri shows me are maltreated—all the good stuff cut down, all the problem trees left to thrive. Stand after stand also is a tangle of tops, the single pattern evident: chaos. "Imagine even the deer trying to get through here," Puuri remarks. The standing trees are small and defective. Of the loggers, Puuri says, "They're fooling the landowner. Take the best of it and the heck with it . . . It's going to be many, many, many years before they have another harvest. You'd have to go a long ways to have a decent tree. There's nothing left—just bad trees. This is really bad forestry, but the average person who drives by here will say this isn't bad. But what it could be, it could be a fabulous forest—what might have been but wasn't. . . . We can do so much better."

Puuri then shows me good timber stands. They look orderly and they look as logical as a family portrait—different ages, different sizes, different shapes. Of course, the skid trails are evident, but, as Puuri says, "Some damage is unavoidable. The trees are down, the roads are messy, but even a good surgical team leaves a trail behind." The key here, he says, is taking out the big, mature trees before they deteriorate like old people; taking out poor-quality trees of any size before they proliferate like bad habits; leaving behind good stuff, young, straight and tall, to thrive like practiced virtues. "This is valuable land when it's in this condition," Puuri says. Such woodlots, he adds, "That's what good forestry is all about." He loves to behold it, and he speaks of it with ardor: "Lovely," he says. "Superb." "Fantastic." "Excellent."

It takes a few years after thinning a stand for the residual trees to develop more crown. When they do, the additional leaf area means the tree catches more sunlight, which turns into more stem growth in both girth and height. Even though encouraging a slow-growing maple stand is a bit like prodding a snail, helping the trees along excites Puuri, so that he says of one stand, "You can almost hear these trees grow," and of another, "You can almost see these trees grow."

Puuri, like virtually every forester I meet, manages his own timber stand with the same resolve he shows at work—not at all like the plumber who fixes everybody's leaky faucets but his own. Puuri's eighty-acre experimental plot is near Toivola. "My favorite place in the world," he says. He loves cutting firewood on warm summer evenings—for "play money," he says, "because I have all kinds of expensive habits, a nice truck and those things." He brings an eclectic vision to managing his woodlot. He has an eye for a hole high up in a dead tree, which he will let stand for a raccoon den. He also has an eye for diversity. He notes that, in the U. P., the dominant maple tree will take over a stand, and edge out other trees. Foresters can encourage that tendency by, say, harvesting white birch because it's a short-lived species. Puuri won't do that. "Isn't it nice to have the color of that white birch in here?" he asks. Right he is. Where the dark maple and oak trunks deepen the shadows of the forest, Puuri's birch stands out like a cresting wave. "I'm pretty proud of how this looks," Puuri says. In one part of his stand, Puuri is fostering red oak. He has transplanted seedlings and "heeled in" some acorns, and he hopes to achieve a stand dominated by red oak. He has a path to some of the imposing oaks blazed with fluorescent orange strips of plastic. Tied on saplings, these trail markers look like garish bow ties. "I let people walk among the giants," he says. "High-quality red oak like this brings a pretty penny," he adds. "They're nice, aren't they? If you like nice old trees, these are nice old trees."

Like marker Jim Johnson, who felled trees with the detached attitude of the professional, Puuri says he is purely practical when it comes to that task. "When it's time to harvest trees," he says, "I have no heart." But this self-assessment is not entirely true. Sometimes Puuri's sentiments override his practicality. He recently looked over a woman's stand of towering white pines near Jacobsville, twenty miles south of Lake Linden. He remembers telling the woman: "God isn't producing any more of these big, big ones all of a sudden—like, those are cut, they're gone. You won't have another one for three hundred-fifty to four hundred years." Puuri says the woman put her arms around him and said, "I like you."

Like the big pine, Puuri's big oak stir awe. One is eighty-one inches around, a beauty, straight as a good conscience. It would make any ordinary logger larcenous. Puuri regularly measures his oak to record growth in diameter, height, and volume. He allows that all of them are mature and "aren't putting on any world-beating growth." In 1991, one measured 25.7 inches in diameter; two years later, 26.0 inches. Puuri says: "It's like getting an old man to grow." Nevertheless, he has made a pact with God to stay for the moment the axe, not to get greedy, to let them grow to thirty inches in diameter as long as they're healthy. Such restraint would be good for all of the U. P.'s hard maple. "We have the best sugar maple in the world," Puuri concludes, "and can return big, big bucks if managed. . . . We could have good forests here for a long, long time—forever, if we take care of them."

Gus Erdmann

THE DAYS TO COME

Part III

"We're for the hereafter."

An old Italian immigrant told me that the secret to a good garden is "dig, dig, dig." Foresters say the secret of a good forest is cut, cut, cut; for the basic tenet of forest management is that people can grow trees better than nature— "faster and better," says Gayne (Gus) Erdmann of Marquette, the biggest city in the U. P., a hundred miles southeast of Houghton. "There is no question about that," he adds. "We can. We know it. But you have to have good foresters to mark it and good loggers to cut it."

He himself feels a kinship with forebears like Aldo Leopold, John Muir, and Gifford Pinchot—all instrumental in establishing national parks, national forests, and wilderness areas. Erdmann says of the legacy of those men, "That's where we're from, eh? That's the kind of forester I am." And he adds of other foresters only in the business for the money, "Some of us are dollar-a-day foresters, and we shouldn't be. Get out of the business if that's what you feel about it."

Similarly, Erdmann says of loggers: "There are two things I look for in a logger—careful and honest. And that combination is hard to find." He has had firsthand experience with what he calls "cut-and-get-out" loggers. Since 1957, his family has had property near Lac du Flambeau in northern Wisconsin. "The logger cut all the good trees and left all the poor trees," Erdmann laments. "Then he sold it. He does that for a living. He's still doing it. Then some fool like us buys it." His land has only three crop trees per acre where there should be a hundred.

199

These matters are on Erdmann's mind when I meet him in late winter at his home in Marquette. Right off, he complains about his weight. He allows that he always puts on pounds in the winter and always sheds them in the spring when he starts running survey lines and marking timber, jobs he loves and is impatient to get back to. "I'm happiest when I'm out there someplace," he says, gesturing towards the outdoors.

Erdmann is retired from the U. S. Forest Service. He studied forestry at universities in Wisconsin and Montana and began his career as a trail builder, fire lookout, and smokejumper out West. He says of parachuting into forest fires: "It was fun. Risky, though." Once he got tangled up in a partner's chute and landed in a tree.

"I thought I'd had it," he says.

"Did your life pass before you?" I ask.

"There isn't any time for that. Things happen too quick."

After those adventures, Erdmann worked six years with the Forest Service in Iowa, and then spent from 1966 to 1989 as a research forester at the North Central States Forest Experiment Station near Marquette. He helped continue work, begun in 1926, that saw foresters apply an array of management practices to different parts of the six-thousand-acre tract, called the Dukes Experimental Forest. The result: what foresters call "the bible" on how to manage northern hardwoods in the Great Lakes states. The chapter and verse of these studies show the ideally managed forest right down to the exact number (320) and size (202 saplings, 65 poles, and 53 saw logs) of trees per acre. The Dukes forest, then, was beginning to show the best way to grow hardwoods while they were simultaneously being felled wholesale elsewhere in the U. P.; for during the early part of the century, 80 percent of Michigan's hardwoods were being clearcut to fuel the state's charcoal industry.

These days, some of the state's hardwoods are being clearcut again, so Erdmann feels a sense of urgency when he applies the Dukes research knowledge to sites that he contracts to mark and when he champions what he and his colleagues say is the best management technique for hard maple: individual tree selection. As he does that, he has one eye on the past, the other eye on the future, and an overall picky attitude about his work. He says of the excesses of the past: "We're beyond that—'Do anything you please.'" He says of good forestry management: "We're for the hereafter." And he says of his own effort: "I'm willing to work for you if you're interested in handing it down to your children. That's the first question I ask: 'Do you care about the future?' If you're not interested, I wouldn't work for you."

As Erdmann tries to manage hardwood stands, he uses different practices, which he calls "treatments." They number four: clearcut, shelterwood cut, diameter cut, and individual tree selection. Erdmann explains each and assesses each with a blend of both the ideal and the practical, from the standpoint of both caretaker and businessman:

The *clearcut* essentially lays the forest flat. The term and the practice are as charged as lightning. People see it and don't like it, Erdmann says, but they also don't understand it. Aspen, a common pulpwood, thrives with a clearcut. Under certain conditions, Erdmann says, clearcutting other hardwoods, including hard maple, is appropriate. It is the most practical treatment for mountainous regions in the Northeast where inordinately expensive operations—including what he calls "ten thousand-dollar logging roads"—are standard. "You cut it and get out," he says. "You're not going to come back for a hundred years." Such a cut in such a place he calls "timber mining." The practice, he says, is appropriate for some of the rocky, hilly, poor sites that are common to Marquette County.

A *shelterwood cut* is kin to a clearcut; if a clearcut is an amputation, a shelterwood cut is an amputation with a prosthesis. Let's say an even-age hardwood stand, the product of a 1930s clearcut, is characterized by a closed forest canopy with little reproduction established beneath it. The understory is overly suppressed and of poor quality—consisting of what some foresters call "craplings." In such a stand, a four-inch sapling might be as old as a twenty-inch tree. If the site is untouched, Erdmann says, "What you see is what you'll get"—namely, more bum trees. So the ameliorating treatment is a shelterwood cut: first, remove the poor understory and open up the crown by 40 percent to allow plentiful regeneration; then, after three or four years, remove the entire overstory in the winter to avoid damage to the new eighteen-inch seedlings. The key here: established regeneration. "We go right down to ground zero," Erdmann says of the technique, which is best for pulp stands, but also can be applied to previously mismanaged timber stands. The downside: a downtime of fifty years before the next pulp cut; a downtime of double that before the next saw log harvest. "That's a long time frame when you're paying taxes on it," Erdmann says.

The third cutting technique, a *diameter cut*, is simple and straightforward: a tree is so big, you cut it; fell every tree that has a minimum diameter of, say, twelve, fourteen or sixteen inches. (A very small diameter, say, five inches, constitutes a clearcut.) The diameter cut is inexpensive; generally, there's no marking, just the logger's practiced eye. But the diameter cut, Erdmann says, ultimately degrades the stand. The practice tends towards harvesting increasingly poorer trees. "You're killing your top horses all the time," Erdmann says.

The fourth and last cutting practice, *individual tree selection*, is the best way to manage hard maple. This process, which takes a long time to get in place, ultimately rewards the landowner with a sustained yield of sawtimber that is of high quality and high value. The practice is based on the classic four-story tree stand that Jim Johnson and Erdmann call "structure." Each generation of tree leads the way for younger trees to follow. Density makes trees grow nicely.

Reaching for light yields clear, straight stems. The forester and the logger periodically enter the stand and thin it. Thinning doesn't increase the rate of growth of the stand; rather it concentrates the growth on fewer, more desirable trees. With each entry, the forester assesses the entire structure, picking the trees to nurture and the trees to fell. The practice steadily upgrades the stand to achieve a constant growing stock of 6,000 board feet of timber per acre. With that stocking, an acre of hard maple will grow 300 board feet of sawtimber per year, in essence, only a few logs. Those few logs are too few to harvest efficiently and economically, so typically the forester and logger put the stand into a ten-year rotation, so that a total of 3,000 board feet per acre, or 120,000 board feet per forty, is cut every ten years.

All of which is not to say that landowners have to manage good sites for sawtimber. "There's nothing wrong with an individual going out and taking their quality sites and managing them for pulpwood," Erdmann says. "They can do it. I don't see any harm in that. They're paying the taxes. If they want to manage for pulpwood, that's okay. They can do better by managing for saw logs, if they were looking to the future further. That's their mistake. I hate to put regulations on what they have to do. I hate to be the guy to say to the small landowner, 'Well, you've got to do this.' I'm a private guy, too, you know. I got land. Just because Gus Erdmann would like to see good-quality saw logs, doesn't mean you have to do that. Pulp on hardwood sites can be good business."

On good sites, though, Erdmann prefers to manage for saw logs and veneer. The three keys to that type of management are a resilient forest, a circumspect marker, and a patient landowner. The northern hardwood forest, Erdmann says, naturally favors hard maple. "This is sugar maple country. That's the easiest thing to manage for. That's the way nature works." The marker, he says, must have the vision to see the forest as it might be. When he fells a mature tree and creates a canopy gap, hundreds of thousands of seedlings will establish themselves on the forest floor below. In thirty years, those seedlings will thin down to five pole-size trees. In one hundred years, those poles will thin down to one twenty-inch-diameter tree. The marker will mark this mature tree for harvesting before it starts deteriorating. And then the marker must envision the long, slow process repeating itself. Meanwhile, the landowner must have the self-discipline to stay the axe until the structure of the stand is ideal or close to ideal, after which it will sustain regular, periodic cuts.

Establishing sustained yield is a job replete with patience. "You gotta wait," Erdmann says. It's also a job replete with judgment, but all too often, he adds, the ultimate judgment is based not on good forestry but on short-term economics. Whether small loggers or big, industrial, private landowners are involved, many are chafing at the bit to take advantage of the unbelievably good 1990s market. Indeed, Erdmann says, loggers in Lower Michigan and Wisconsin have "raped" hardwood stands, hoodwinking owners who don't realize the value of their timber or the proper way to grow and harvest it.

"I hope we don't get that kind of progress up here," he adds.

As for industrial landowners, their foresters are "no longer in positions of power," Erdmann says. "Accountants are running the forests, and that's sad. They think in the short-term. Last quarter's profit drives them all." The arithmetic can be instructive. At early 1996 prices, an acre of growing stock could bring in $2,400, while an acre of annual yield could bring in $120. The result: industrial forest owners are cutting down the growing stock while the market is good—in essence, liquidating the forests. "It almost brings tears to my eyes," Erdmann says. Liquidating essentially means cutting all the sawtimber, leaving the less valuable pulpwood. Liquidating borders on clearcutting, and a clearcut forest takes a hundred years to regenerate and produce merchantable timber. Thus, with liquidating, an acre could bring in $2,400 now and not a dollar more for the next hundred years; with selective cutting, the same acre could bring in $120 or more every year for the next hundred years—and produce a total income of $12,000. Nevertheless, landowners are increasingly seduced by the $2,400 in hand versus the $12,000 in the bush over the long haul.

Erdmann acknowledges these economic realities, but some of the pressure on the resource, he says, is offset by a tax break that the state gives to landowners in return for their having a forest management plan. "We've come forward a little." He also is encouraged by the efforts of Dick Bolen, Carl Puuri, and the Forest Improvement District to educate the small private landowners, who own a quarter of the nearly five million acres of forest in the western U. P., equal to the acreage owned by industrial firms. Educating those owners is vital, Erdmann says, because most people don't know a good tree from a bad one. "A tree is a tree is a tree to most people," he says.

Speaking here is a practical forester who is in the business of growing trees, but who can look at a dead tree lying on the ground and see value—for a salamander to live under, a porcupine to crawl in, or a partridge to drum on. Speaking here is an idealist, and, as he talks, Erdmann picks up a copy of *A Sand County Almanac*, Aldo Leopold's classic book on people and nature, and says, "My feelings are more along this guy's line," he says. "We really should do what's right for the land, and good for the people. We're for the hereafter. We've got to leave something here. We shouldn't be cut-out-and-get-out anymore. We gotta live here." Speaking here is a dreamer. He sees the timber owner nurturing the forest with sound silviculture: "Our stuff is so valuable right now, my god! if we could get the small woodland owner to manage his stuff, we're in business for a long, long time." Concluding, he says, "Good management is reward in the future, eh?—in perpetuity."

One firm logging in the U. P. has raised Erdmann's and others' ire. They say that the outfit used to be a family business with a reputation for fining loggers who *cut* an unmarked tree. Now the company has a foreign owner—one of two such firms in the western U. P.—is cutting heavy, and has a reputation for penalizing loggers who *miss* cutting a tree. "They're doing the worst possible thing you can do," Erdmann says. "They're not clearcutting. They're doing worse than that. They're going backwards. They don't have established regeneration. They're taking the future right out of the stands. . . . What they want to do with the land is cut the product off it and turn around and sell it. . . . Cut it off, and the person that inherits it, he's got what my family has, 1957 to now, eh? We got nothing." He adds as an afterthought, "except taxes."

Albin Jacobson

THE DAYS TO COME

Part IV

"I don't think our forest deserves superlatives."

A Finnish proverb says, "Do good with your own money, not your neighbor's."

Albin Jacobson of Baraga brings that charge about minding one's own store to bear on the economics of commercial forestry. The big woodlands owner says that you need tough bark to be in the timber business these days ("Some people are quick to condemn anyone not operating according to the perceived proper way"), and he has this message for busybodies who would presume to tell him how to manage his timber: "Well, my land—I take care of my own land. You don't know near as much about that land as I do. Don't come in and tell me how to harvest my land."

Jacobson, who says that the growers of pulp and sawtimber are "locked in mortal combat," is an apologist for industrial forestry—even if it means liquidating forests. The firms that are liquidating, he avows, have no workable alternative.

Speaking is a man who, says Paul Clisch, came by his forestry "from the roots up." Jacobson's father was a forester, and he worked in the South, the Philippines, and the Pacific Northwest before returning in 1963 to his native Michigan, and the shores of Keweenaw Bay.

The younger Jacobson was born in Arkansas.

What did you do there? I ask during our first visit.

"I just grew up," he says.

How old were you when you moved to Baraga? I ask.

209

"I would be twenty-one or twenty-three, in that neighborhood," Jacobson says.

Occasional vague remarks are characteristic of him, as are strong opinions.

His father, who helped establish the Seney Wildlife Refuge in the central U. P., bought twenty-four thousand acres, nearly forty sections, in Baraga County east of L'Anse, and he and his son founded All-Wood, Inc. The younger Jacobson still owns, manages, and logs twelve thousand acres, which is stocked mostly with hard maple, hemlock, and birch, but he has shut down the Baraga sawmill that he and his father had bought and rebuilt, and that they operated from 1968 to 1986.

"There's been a sawmill on this site for over a hundred years," Jacobson says. "We never made any real money sawmilling. There were a couple of fat periods, but basically sawmilling has always been marginal for us. Why? Well, if you track—except for a recent period—the price of hardwood lumber against other commodities, you'll find that it just didn't keep pace during the time we had it."

Jacobson shows me his idle mill. It is dark, dank, and dusty. Just inside the entrance, long loops of big-toothed band saw blades lie about, curled up like ribbons. These days they cut through nothing but time.

When we return to his office, Jacobson folds his tall, lean body into a squeaky chair. He looks the picture of tiredness, for he often takes off his glasses, rubs his eyes slowly, and runs a hand through an unruly pompadour. He works long hours. "When the brains leak, the body has to suffer," he says. Samples of paneling, made of wood from all over the Americas, hang on one wall of his office. A calendar with a picture of a Japanese woman in a colorful kimono—a gift from Takano Kimpara—hangs near a sign by the entrance door, which swings out. The sign reads, "Pull hard, push easy." On my first visit, when leaving, I pull.

Jacobson, says Paul Clisch, "goes to his office every day, reads the *Wall Street Journal*, and thinks." Jacobson certainly has a thoughtful bearing—and a picturesque manner of speech. He says, for instance, about one little subject that weaves through our discussion: "This is a thread—not a spider web, and not a cable."

He is a renegade voice among the many foresters I talk to. Is the western U. P.'s timber exceptional? "I don't think our forest deserves superlatives," he answers. "Our forest has a huge advantage in that it is self-propagating. You just have to get out of the way, and it'll grow itself back." Another "phenomenal advantage" that the region's forests have, he adds, is limited exposure to natural catastrophe, such as fire, tornadoes or ice storms. With these considerations in mind, Jacobson asks, "Is our resource so vast and special?" His answer: "I don't think so. It's got some big advantages, but unique in all the world? Naw."

Jacobson is an utter realist about the timber business. Preserving wilderness is a luxury, he says; working the forest is a necessity. "If you don't make a profit," he says, "you're finished. If I can make a profit, a dirty ol' profit, managing that forest without destroying it, then you have a use of that land that is justified. You have to utilize the forest for the benefit of mankind." Many environmental concerns are also a luxury, he declares. "If we enter a major economic turndown or depression, we're going to find this environmental stuff is going to be very secondary to somebody making a living and having something to eat."

Jacobson may be opinionated, but he's not slapdash. He talks mostly about what he knows. I ask him, for instance, in what condition are the area's timber stands that are owned by small private landowners. "I'm not there to look [and] I pretty much mind my own business," he answers.

Here is a person who is at once strong-minded and dispassionate, and I remark on that. "I call that being professional," he says, adding: "The forester doesn't look at the clearcut forty or one-sixty as a huge devastation of the resource. He comes back and looks at it in three or four years—he's trained that way—and sees that land as fully stocked, fully occupied by a young stand. . . . A generation in the life of a forest might be several generations in the life of foresters, and so they look at it and say, 'Well, you know, this thing goes on long after we have passed.' So they take a longer view, a view of the forest as a complete cycle—from seedling to either being harvested or falling down and dying."

Paul Clisch also says, "If anyone overmanages his timber, it's Albin Jacobson." Indeed, Jacobson, like his father before him, has steadfastly applied individual tree selection to his forest, even when he lost money doing so. Still, he understands why others liquidate their forests, and he says that hard maple's problems derive from the very nature of the tree.

He notes that a forest grows a fixed weight of wood fiber a year, and the forester decides which trees to put the weight on. Ideally, the forester nurtures the forest until he achieves the 5 percent growth rate of the Dukes forest, which is the owner's gross return on his investment—the income on his standing timber, before costs and not including the end value of the product. "Any corporation that owns its timberland strictly as an investment would not be satisfied with the rate of return that the forest growth gives them," Jacobson says. Bigger and faster money can be made, though, because, while a mature tree grows only in girth, a juvenile tree grows faster—and in both girth and height—so it pays to grow small trees. "What you see there is the fundamental driver of the liquidation of the forests up here," Jacobson says. "It's not greedy timber barons. It's simply slow growth of the forest, an innate characteristic of the land. . . . If you talk to anyone in the hardwood industry, they'll tell you how the log being processed is shrinking, shrinking, shrinking, because no one can afford to grow big trees anymore. The better than average return from a tree growing in height as well as girth encourages the landowner to hold his resources in smaller trees . . . and so encourages the liquidation of the larger trees."

"What does that do to the resource?" he asks rhetorically. "Well, I don't know that it necessarily has a detrimental effect. I really don't know. People have a romance to see big trees, and I enjoy that, too, but if you have a forest of small trees, is that necessarily bad? I don't think so. I don't see any problem with the forest reproducing itself here. Everywhere you go in the U. P., it's green." And, he adds, young trees, as with people, generally are more healthy than the old.

But people's emotions, Jacobson says, often cloud the issue of forest management, which he sees moving into the arena of property rights. "Does a person have a right to do what he wants to his property," Jacobson asks, "or does society, perhaps because of its emotional prejudices, have a right to restrict what a person does with his land? Obviously I'd have a bias on that, but it's not for me to decide. I'm one vote." Were he to cast it, his vote would be hands off.

Against this backdrop of emotion and economics, Jacobson eschews criticism of the company that is liquidating its timber stands.

"They have to liquidate," he says.

"Why?"

"If they didn't liquidate, they'd run out of money. They have no choice, in my opinion. They have absolutely no choice."

Like Erdmann's, Jacobson's arithmetic also is simple. The company's land, unlike the ideal Dukes forest, more than likely had a growing stock like his own (5,000 board feet per acre) with a yield like his own (150 board feet per acre per year), or 3 percent instead of 5. The company's sawmill cost $2 million, and it has an appetite of 10 million board feet a year. With the forest yielding only 3 percent, the company needs perhaps $40 million invested in timberlands to supply the sawmill. "So you've got a two-million-dollar dog wagging a forty-million-dollar tail," Jacobson says, "and the tail is giving you three percent, and you told the guy running that plant, 'I want you to make me a good return,' and he's stuck before he starts. . . . He can't do it. He can't do it. . . . You need an astronomical amount of acres, and each acre carries a stumpage value, as well as a land value." A $2-million sawmill can't generate sales to justify that kind of investment in timberlands, Jacobson says. "No way." So, he says of the company: "They cut their timber, didn't they? And everybody else is cutting their timber. That's why."

Another issue in the timber business in northern Michigan revolves around a paper mill, "one of the most capital-intensive industries in the world," Jacobson says. He offers some numbers about their finances that "blow your mind when you start pushing them around with your pencil." Paper companies, he says, are liquidating their growing stock and turning the land over to growing, essentially, brush— that is, fiber for the pulp mill. The liquidation generates immediate income, frees the capital invested, and lowers the book value of the land, in terms of dollars invested in timber, down to a few dollars per acre. When a $300-million mill owner does that to its valuable timber holdings, Jacobson says, "You have a big dog wagging a modest tail." The essence of the matter: a $300-million paper mill can carry a lot of $10-an-acre land, but a $2-million sawmill can't carry a lot of, at best, $2400-an-acre land. "That's a driver for liquidation," Jacobson says.

However compelling the numbers, Jacobson is only a devil's advocate. In reality, he would say, "Don't do as I say, do as I do," for he manages his own land with individual tree selection, logs his tract on a five- to seven-year rotation, and cuts about two thousand acres a year. His average stocking and yield are less than in the Dukes stand, he says, because larger tracts are harder to manage and have what he calls "inevitable void"—by which he means swamps, rocky areas, roads—and bad decisions; for he adds, "We're capable of making mistakes, too, you know, so you have losses."

He's making money, though, and adding to the black of his ledger is what Jacobson calls the "recreational value" of his land. He leases quarter-sections to outdoor enthusiasts—to hunters, for instance, who build hunting camps. The cost: the property taxes (about four bucks an acre), plus 10 percent. The scheme has worked out well, but even Jacobson is a bit surprised, for when he studied forestry in college, he grappled with the idea of the recreational economics of the forest—for instance, the worth of a wilderness area. It was such a slippery concept that he wrote the idea off as "total nonsense." Nowadays, with the pressures on timberland more intense, he has learned otherwise with his leasing program. All of his land is close to being fully used, and, after several years of tinkering with the setup, he has determined the recreational value of his land: 15–20 percent of the value of his timber growth. One of the selling points of his program: leasees prefer nice timber stands to pulp or brush stands.

Jacobson continues to log on the leased parcels. On all of his land, he restrains himself from liquidating and strives to grow trees that will yield large saw logs and veneer, two to three logs to a stem. Over the long haul, high stumpage prices, plus the fact that good big timber is being liquidated and getting "rarer and rarer," means Jacobson's trees have become extraordinarily valuable. He is not so brash as to say their value will continue to climb. "That's the concept we're working under," he says, "and in the last six or eight years, that's true. But I can't say that will continue to be true in the next ten years. I'm not clairvoyant. I'd like to think I'm doing something very smart and right, but I can't prove it."

Has he made money so far?

The last ten years, with the unusually good market, have been a welcome boon. "Now," he says, "unless you don't mind the store at all, you're going to make money." Still, he is not making nearly as much money as he could if he liquidated. And if the last ten years have been profitable, the first ten years that he and his dad worked their timber stand were an unequivocally losing proposition, and the second ten years were only marginal—all of which prompts Jacobson to muse, once again, about the economics of the timber business. "In most financial circles, fifteen years is equivalent to infinity," he says. "If you went to Wall Street and said you were willing to invest and lose money for fifteen years, they'd call you crazy—a lunatic."

But that's what he and his father did. What did it take to weather those early years of red ink that was as pronounced as the heartwood of a hard maple tree?

"Youth and idealism and faith," Jacobson says offhand. "Staying power." Plus, and perhaps most importantly, "a family fortune." He notes that his father continually pumped money into the company to weather those bleak first ten years. Jacobson says his father was "the real brains" of All-Wood, but he died in 1981—several years before his vision of the value of a sawtimber and veneer forest was confirmed. Jacobson says that he has paid his dues, too—by not breaking faith with his dad, by steadfastly resisting the strong inducement to liquidate, a temptation he says he wrestled with "intensely."

Today, Jacobson says All-Wood's position in the marketplace is marked by duality. The more the U. P. forests are converted to pulp stands, the more his sawtimber is worth. But the more his growing stock is worth, the more "pitiful" his 3-percent income seems. Liquidation becomes more attractive each year, especially with one ogre lurking over his shoulder. "The taxman knocketh," Jacobson says. Inheritance taxes, he explains, "make it essentially impossible for me to turn my timberland over to my heirs. Or, if I turn it over to my heirs, my heirs almost categorically have to liquidate the timber to settle the estate. I have no solution to that problem. And that bothers me. That's kind of sad. But that's society's problem, not Albin's problem."

"In the meantime?"

"In the meantime, I do what I've been doing. I have nothing else I'd rather be doing."

So far, so good. He has sold half of the land that his father bought. "I had quality timber when quality timber was hot," he says. He made a handsome profit. He did *not* liquidate before he sold, but most of the new owners have. He sold that much of his land because of college expenses for his children, the imminence of old age, and other such everyday considerations. "Love and kisses won't get it anymore," he says. "You have to have the cash."

Jacobson also sold some of his land partly as a hedge against another major depression, which he smells in the wind. "We are overdue," he says, adding that, if one hits and he hasn't liquidated, "Then you start saying, 'If only . . .'" In that eventuality, he can just picture himself saying, "Yes, I made a small fortune in the forestry business. Unfortunately, I started with a rather large one." In any event, by selling half of his holdings now, he hopes to avoid what he calls "a distress sale" later. He figures that he now has enough money to ride out a downswing in the maple market or in the entire economy—enough, he adds, "to shut down and not operate a long time, if necessary. . . . I can sweat out an awful lot of heat."

As for others, Jacobson sees a cold reality: the likelihood of liquidation. He sees the FID teaching landowners about the value of their maple, but he also sees the landowners succumbing to the temptations of the bull market. Says Jacobson, in his no-nonsense way, about any one of these people: "Just about the time we get him educated, he wises up."

Bob Darling

THE DAYS TO COME

Part V
"Okay, Erin, we're going to write a song now."

Like startled beavers slapping their tails on the water, some residents of the Keweenaw Peninsula recently sounded their warning: somebody wants to build a pulp and paper mill in the area. The possibility had them seeing red—the Keweenaw's forests bleeding to death from rapacious logging.

Like song birds bringing in a new day, other residents of the Keweenaw Peninsula, at the same time, sounded a clarion call: somebody wants to build a pulp and paper mill in the area. The possibility had them seeing green—the Keweenaw's forests thriving from much-needed heavier logging.

The two messages were heard across the length and breadth of the peninsula, from Silver City to Copper Harbor, from Keweenaw Bay to Manitou Island. Some of the people sounding the warning organized as FOLK, Friends of the Land of the Keweenaw; some of the people sounding the clarion organized as PULP, People Unopposed to Local Progress. Their ensuing debate involved the standard polemics when anybody anywhere debates local development: the price of growth, the cost of no growth; the appropriate use, and the abuse, of natural resources.

James River Corp., of Virginia, kindled the debate in 1989 when it announced plans to build a pulp and paper mill, at a cost of $1.5 billion, near Arnheim, which is about eight miles south of Chassell, in northern Baraga County. The mill would be the fourth in the U. P. Not long after the announcement, a resident of the paper mill town of Quinnesec said the issue would divide the community. He was right. People who opposed the mill feared that the peninsula's waters, air, and forests all would be degraded; people who supported the mill said that the forests would be rid of junk timber and upgraded, and jobs would be created: 2,000 construction jobs, 500 mill jobs, and 1,000 jobs in the woods to feed the mill with pulp—all this in a county with fewer than 9,000 residents who at the time were enduring 9 percent unemployment.

"How can you put a price on something priceless?" one mill opponent said in trying to assess the worth of the Keweenaw's relatively unspoiled environment.

A mill proponent countered: "The worst pollution affecting mankind is unemployment."

The idea, then, of 150 logging trucks loaded with pulp rumbling down the highways every day to feed the mill was a blight to some, a thing of beauty to others.

James River wanted to build "a world-class" mill on the banks of the Sturgeon River, upstream from Chassell Bay, which is part of Portage Lake, which connects to Lake Superior. The company planned to make white paper with a bleaching process that dumps dioxin, an affliction to environmentalists and some scientists, into the watershed. FOLK, which organized almost immediately after plans for the mill became public, opposed the mill primarily because of the spectre of polluting the Keweenaw's waters. Thus, while mill proponents said that Michigan had environmental safeguards tough enough to minimize pollution, mill opponents said the technology *not* to pollute badly simply doesn't exist. Air pollution also was a major concern for FOLK. The stench of rotten eggs goes hand-in-hand with paper mills, so FOLK went nose-to-nose with James River and its supporters over that issue. Lastly, forests—"the blessed and untouched forests of the Keweenaw," one letter writer dreamed—mattered greatly in the debate. Concerns about them ranged from the philosophical to the utilitarian. One mill opponent wrote that the forests should be managed "for the delight of beavers, deer, owls, trees . . . and the poet in every human." Another opponent feared that forests would be overcut or clearcut to supply the mill.

FOLK members, with great effort, pounded away at getting an important part of its message across to the public, namely, that they supported a timber-based economy for the Keweenaw, but not at the high price of environmental damage. Rather, they favored small industries like furniture and veneer mills that they said have a less rapacious appetite for northern hardwoods. They urged Keweenaw residents to forego the certainty of heavy industry in return for the possibility, ten years or so down the road, of new and clean technology to manufacture wood composite panels out of the same kinds of trash trees that would go into making paper.

For its efforts, FOLK and other mill opponents were branded with aspersions, albeit mild ones. Mill opponents did a bit of name-calling in return, and, as a group, they invoked all manner of images and notions to advance their viewpoint. Must tea bags, they asked, be so white that bleachcraft paper mills are needed?

The paper mill issue became known as "The Battle of Arnheim." One side made economics and jobs the paramount concerns; the other side beauty and cleanliness and the quality of life in the Keweenaw. Of the latter persuasion was a homeowner from rolling land about six miles west of Arnheim, near the Otter River. He lived in Elo, which is a Finnish word for *alive*. In Elo there is a store, a closed-down church, and a cemetery. The man wrote about jobs versus environment: "I don't believe in tradeoffs. I believe in wholesomeness. I live in Elo country and I love it."

Six months after it first broached its plans, James River pulled back from the mill proposal. Officials said the decision was based on market considerations, not local opposition. In the aftermath, the debate continued over the future possibility of a mill elsewhere in the Keweenaw, perhaps Ontonagon, on the other side of the peninsula, some said. PULP helped keep the issue alive by staging a straw vote in Baraga County. Nearly three thousand people voted. The pros beat the cons by 159 votes. Both sides claimed victory.

The imminence of a pulp mill in the Keweenaw has receded, but the idea of having one is still a possibility and continues to stir passions and opinions. One person who is part of the mix is Bob Darling, president of FOLK. He opposed the pulp mill when it surfaced in both Baraga and Ontonagon counties, and he still does six years later. Darling is forty-five. His involvement in FOLK is his first in an environmental group, but he says the seeds of his interest were sown when he was a boy.

He used to spend summers on sixty acres of bottom land just outside Chassell. His grandparents, from Chicago, bought the place and retired there in 1954. At the time, the land was wild, with no yard—mostly just weeds, brush, small trees, and swamp. His grandparents nurtured a farm out of some of the acreage and grew blueberries, strawberries, asparagus, and other vegetables, all of which they sold commercially. Darling says that his grandparents were environmentalists before that term gained currency, and they reused, reduced, and recycled before those terms were the jargon of conservation. "It was a way of life," he says. His grandparents are dead, but their ways rubbed off on Darling. "You're a product of what you're taught as a kid," he says, "and those were values that to me were just drilled in—'Just don't waste things. It's not good sense.'"

I talk to Darling twice, the first time in a coffee shop, the second at his family's farm. I drive to the farm on the first day of August. People are picking thimbleberries along the roadsides, and the maples are starting to show their fall color. A long, dirt driveway leads into the farm and ends in a large, grassy yard lorded over by trees that his grandparents planted: red pine, jack pine, white pine, poplar, and spruce. The driveway ends amid a cluster of buildings, including a big barn that is made partly of hewed and chinked timbers. All around, chickens pick a little and talk a little. When we enter the house, two lazy dogs settle on the front stoop, and a sleepy-eyed cat curls up by the window in the front porch. Darling and I sit and talk at an old, oak kitchen table. At one point in our conversation, a heron flies so low through the front yard that it startles the chickens, which sense perhaps the shadow of a hawk and fuss noisily. The heron lives by a pond behind the house. A year ago, Darling bought a hundred nine-inch speckled trout for ninety cents each and planted them in the pond. Now there are few, if any, left. Darling's neighbor, a fur trapper, suspects otter and mink have fattened up at Darling's expense.

While Darling and I talk, the peace of the countryside is hammered away by his father, who is sixty-five, and a nephew, who is fifteen. The two are remodeling a grain barn-turned-cabin. The sight of the old man and young boy working together prompts Darling to talk about what he deems is crucial to growing up right: family involvement, self-reliance, and respect for life—matters he learned as a child when he summered on the farm. He says that he didn't have to be entertained during those halcyon days of his youth. He picked berries, imagined a stick to be a musket, explored the big swamp on the back twenty, and fished. The Pike River and a no-name creek course through the farm, and he caught trout from both when he was young. It's been said that trout are to water what canaries are to coal mines: if trout thrive, the water is cold, clean, and well-oxygenated; if trout leave or die, the water has become warmer, turbid, and less oxygenated. These days, there are very few trout—"nowhere near what there used to be"—in Darling's waters, mostly just chubs and suckers. "It makes me sad," he says. Still, good memories abide: wily brown trout in the water, raucous Canadian geese in the sky, ranging black bear on the land. All of it, he says now, was "an adventure"—and a lesson, too, in the worth of living things, and in the value of making his own fun.

Darling returned for good to the haunts of his youth in 1980, the same year he started planting trees commercially, which he did for six years. It was seasonal work and migrant work. He started in the winter in the South and finished in the spring in the U. P. He planted in Mississippi, Arkansas, Illinois, Wisconsin, and northern Michigan. He worked for five paper companies, Georgia Pacific, Mead, International Paper, Potlatch, and Weyerhauser. He also worked for the U. S. Forest Service on the DeSoto, Ouachita, Shawnee, Nicolet, Chequamegan, Ottawa, and Hiawatha national forests. The pay was good, the work hard. On a really good day, he could make two hundred dollars. He planted an average of two thousand seedlings a day. It was a dirty job, even hazardous; for, before planting, paper companies sometimes treated sites with insecticides and pesticides. Darling and his cohorts—all college graduates—suffered bad headaches from their first encounter with chemicals. They also knew of a crew of Mexicans that went partially blind for three days on another site. And they saw a crewman's dog sniff around all day on yet another site and lose most of its nose. "It looked like it melted," Darling says. Shortly after they started planting, then, he and his group swore off working on chemically treated sites—one of the first evironmentally oriented decisions in his life.

Darling first joined FOLK because James River planned to locate the proposed mill on the Sturgeon River, where he likes to fish. His favorite spot is near his house, where the meandering Sturgeon River Sloughs, which look like a muskrat's maze, meet Chassell Bay on Portage Lake. Those waters, Darling says, have always been "kind of sacred ground to my family." He says he did not savor "the idea of catching my fish out of pulp mill effluent."

Once with FOLK, Darling brought his tree-planting experience to bear on the mill issue. One thing he witnessed from the work is what he calls a "pulp mill mentality." The term sums up what he fears is the inevitable offshoot of a pulp mill: economic pressure to grow a pulp forest. A mill, he maintains, causes a huge demand for low-quality timber that could result in the planting and growing of low-quality trees. In the Keweenaw, he fears, there would be "intense pressure" to overcut hardwoods and convert fine hardwood stands into fast-growing pulp stands simply because that's what a pulp mill needs. He's seen that scenario unfold elsewhere, and he is particularly vexed when it occurs on public land, which comprises 34 percent of the forest in the western U. P. and which, Darling argues, should not be managed in a way that subsidizes commercial ventures.

Even that concern among members of FOLK, though, was overshadowed by another. Darling, who was on the group's steering committee from its inception and is now the third president, says that the membership, which reached seven hundred at the height of the mill debate, decided early on that it didn't want to be just anti-mill; it wanted to stand *for* something, too. One of its imperatives: the Keweenaw's maple is unique; therefore, it should be used to make products that are unique. Paper, Darling says, can be made just about anywhere from just about any old tree. The Keweenaw competing with other parts of the country in the paper business, he says, is as illogical as Wyoming competing with Wisconsin for the Chicago dairy market.

Darling believes growth on the peninsula is inevitable. "This is a desirable place to live. If you accept global warming as being in our future, that will make this more of a desirable place to live. . . . If you accept the fact that good water is getting scarcer and scarcer, there's going to be an attraction toward the Great Lakes and probably up here. So I think as far as planning and growth, we need to be aware that it's going to happen up here, and we need to allow for that and plan for it." Otherwise, Darling says, the area's attractions could be the seeds of its demise.

Darling acknowledges that "we're all economic animals," but he argues that the economic value of the forests needs to be balanced with the ecological value of the forests. He sees the forests not just as a crop to produce timber and wood products, but also as a place for sustaining wildlife and other natural bounty and beauty. "That's where, as human beings, we draw peace from," he says.

The idea of nature as a stimulus to the spirit is ingrained in Darling's life, including in his work as a counselor of teenagers in trouble. "I deal with kids now who can't go out and entertain themselves unless they have a Nintendo or something," he says. "They can't use their imaginations and just go out in the woods." These youths are lesser souls for that inability, Darling believes. Better, he says, to be as he was— excited about catching a trout, but a bit sad, too, because something died.

Darling occasionally takes some of his juvenile charges to his farm—"to visit my chickens and stuff like that." He also takes them swimming in the Sturgeon River, into the bush at Christmastime to get a Christmas tree and boughs for wreaths, and out in his boat to fish on Chassell Bay. All of it is "pretty much a new experience for them," he says. "... They've never been out in a boat fishing, especially the girls. But boys, too. I have to teach them how to bait hooks and how to take the fish off." He does all that extracurricular stuff not for some high-minded counseling opportunity— just to show the youngsters how to enjoy themselves.

Against the backdrop of the troubled youngsters he works with, Darling's niece, Erin, stands out in sharp silhouette. As Darling was able to do in his youth, she can entertain herself on the farm, and nature expands her world. Her best friend is a hound dog, and one of her favorite pastimes is catching frogs, then setting them free all at once. "Quite a sight," Darling says—seeing fifty frogs hopping away all at once. "It doesn't cost any money and unfortunately doesn't contribute to the Gross National Product, but it's appropriate activities for kids to go through. That's using your imagination for something. Does money factor in? No. But it's growth. It's growth for a little kid to learn how to occupy their time in appropriate ways, using only their imaginations, without spending money."

One way that he himself does that is through music. He is a musician, a singer, and a song writer, and he often uses nature for inspiration. Humankind, he says, is the only life form on the planet that has emotional and spiritual needs beyond physical survival, and nature has inestimable value in meeting those needs.

He recalls once when his niece—she of hound dog and frog fame—was five. One day they sat together on the porch. Darling strummed his guitar and said:

"Okay, Erin, we're going to write a song now."

"What about, Uncle Bob?"

"How about the clouds?"

Darling started playing and Erin started singing.

"Just like that. Singing about the clouds. She didn't know it was hard to write a song. That's what creativity is—tapping into that innocence that everybody's kind of born with." He has written and played music for years—folk, bluegrass, and rock. Now he and the Bayside Boys perform "whatever pleases us." He can play the guitar, bass, drums, mandolin, and violin. Not really well, he says—"good enough to get the job done . . . but keep it simple." So, in music circles, he is called "semi-good Bob Darling." In FOLK, though, he certainly isn't called half-hearted Bob Darling, for his convictions are as deep as a taproot. "We don't want to live on this planet alone," he says simply. So he wants forests to be there for not only lumberjacks, but also for beauty and beast and biodiversity. To those who doubt the worth of that dream, he says in a song he wrote:

So take my hand and walk with me,
and we'll listen to the wisdom of the trees.

———

Paul Clisch would say that Bob Darling is talking rot. Clisch believes the Keweenaw needs a pulp mill. He says the area needs the jobs, and the forest needs to be rid of the junk trees. "You can't manage your land if you can't sell the pulp," Clisch says. Environmentalists gall him. He believes they derailed the mill, and he says of them, "How many jobs did they create since then? You don't see one job. They could care less. If you never cut a tree, they would be happy. They want me to grow trees so they can look at 'em, and that's a dirty crying shame. It's sad, really. What good is something you can't use?"

Some observers believe the western U. P. could support three more pulp mills without adversely affecting the forests. Along that line, Albin Jacobson says that no matter how intensively he manages his timber, at least 30 percent of his growth will always be pulpwood. The pulp issue used to stick in the craw of this normally composed and pensive man. "I used to storm at the paper companies," he remembers—not so much for prices, because sometimes the prices were very good. Rather, Jacobson says, he would get upset about the length of the contracts for pulp: most lasted only a few months, not long enough to do more than begin to write off the capital investment in machinery. It was a feast and famine situation that he tired of.

"So what do you do with your pulp?" I ask.

"Cut it down and let it rot," he says.

But if the Clisches and Jacobsons and Darlings of the world differ on the mill issue, they share an attitude: they all have the future in mind, and they say that they are concerned about leaving behind good forests for their children and their grandchildren. So they make it their business to worry about their forests. It reminds me of two brothers I knew in Colorado. Both were in their thirties and lived with their mother. The younger one took to staying the night at his girlfriend's. The older one told me that he was all set to have a talk with his younger brother.

228

"Is that your business?" I asked.

He told me when his brother didn't come home at night, his mother got real upset. When his mother got real upset, the apple pies weren't as good. When the apple pies weren't as good, that got him real upset. So he had to have a talk with his brother.

A similar scenario plays itself out in northern Michigan's forests. Lots of people have a hand in it; it's everybody's business, but their goals differ, and, in any event, their goals are elusive. Big tree or small, pulp or veneer, clearcutting or individual tree selection, paper mill or no—all of it is hard to grasp and pin down. Paul Clisch sums the situation up well. When I spent the day with him, he cut down some basswood, which grows in clumps. The one he was cutting had three stems. He pointed out that each stem was a different size and had a different-sized heart. He said of the basswood—and perhaps of much more in life—"Everything's different, and it's the same tree."

Whichever way the pulp mill issue washes out, whichever charlatan logger or steward of nature prevails, the debate over the forests of the Keweenaw might become even more critical than it already is. One forest researcher, in an intriguing but controversial bit of statistical analysis and speculation, says that if global warming occurs, hard maple might die throughout its entire range—except for one spot where Lake Superior keeps the climate relatively cool: the Keweenaw Peninsula.

In the meantime, the forests are still here, and so are the people concerned about them. Their collective voice about how to use this exceptional but ailing resource echoes traffic signs I saw years ago in Montana: "Speed Limit," they read, "Reasonable and Prudent."

The author and Paul Clisch

EPILOGUE

Fred Aho died in August of 1995.

Swede Intermill died in January of 1994.

Michael Lorence of Northern Hardwoods continues his travels to places like England, Finland, Russia, Italy, and Germany; and he says he is just following the trails of northern Michigan maple. "Our wood's been all over the world," he says. "This area has been associated with fine timber for years."

Bad knees retired Jim Johnson in October of 1994. "Forty years walking in the bush—that's about the roughest walking you can get," he says.

John Porkka has sold his truck for $28,000, bought a new one for $138,000, and bought a slasher for $40,000. His production is up, but he's not certain how much more money he's making. "I should figure it out right to the penny, so I would know myself," he says. "Maybe I'd quit." Meanwhile, John has learned why he's feeling poorly: he has tick-borne Lyme disease.

Henry Clouthier has dropped about seventy pounds of brawn but none of his spunk. When a friend and I go to photograph him, he says, "I got a two-dollar coat rack that could take a better picture than that outfit you got."

Horner Flooring once was bedeviled by fire; more recently it was threatened by poor management. John Hamar sold the company in 1990. The firm that bought Horner used up the company's operating funds, cash reserves, and lumber inventories. The company began losing orders and customers, and, Doug Hamar says, was within a month or two of shutting down. The Hamar family, with mill employees lobbying on its behalf, bought back the company in

231

1993. In 1995, Doug Hamar summed up the situation: "We're not out of the woods yet."

John and Barbara Clark now have electrical service to their cabin. "It's a lot easier to vacuum," Barbara says.

As he intended, Takano Kimpara has traveled to New York State and Maine looking for bird's-eye. The maple in New York was "so and so," he says. In Maine, which he last visited over a decade ago, the maple was better, but it wasn't as white or as dense as northern Michigan's, and it had more mineral stain. He says he was "just looking" and bought nothing. Kimpara also has traveled to Russia and was amazed at vast virgin forests without a road in them. And he has joined a consortium that opened an office in Shanghai to pave the way for the group's doing business in mainland China.

Not long ago, Paul Clisch contracted to cut some of the biggest maple he's ever seen in northern Michigan. He showed me the stand. The two of us, arms outstretched, could barely encircle one tree that we figure is about eleven feet or more around.

*   *   *